Travelin' with the Poor Boy from Becks Creek

JAMES P. CREEKMORE

Copyright © 2016 James P. Creekmore

Polished Stone Publishing
P. O. Box 2202
Sebastopol, CA 95473

All rights reserved. This book may not be reproduced in whole or in part without written permission from the publisher, except by a reviewer who may quote brief passages in a review; nor may any part of this book be reproduced, stored in a retrieval system, or transmitted in any form or by any means electronic, mechanical, photocopying, recording, or other without written permission from the publisher.

Cover design and interior layout by Lesley Thornton-Raymond

ISBN: 978-0-9980976-1-9

First Printing
Printed in the USA on acid free paper
10 9 8 7 6 5 4 3 2 1

DEDICATION

I dedicate this book to my precious wife Jeanie, and to my wonderful mother who was so devoted to the family. Mother and Jeanie, alike in so many ways, were dearly loved by all. Both their morals and family values have helped us all in raising our families during the hardships and challenges we faced throughout our lives.

As I think back, I realize how fortunate I was to find such a beautiful and precious woman as Jean Standard who became my lifetime companion.

Once I met her and we started dating, none of the pretty girls (or I should say young women) I had dated since high school were forgotten, but they were no longer that important to me.

Jean Standard was my one and only.

I have no regrets of the wonderful years we spent together with our wonderful sons.

I also dedicate this book to my caring brothers and sisters, and especially to our older brother Bill who kept the family laughing in the roughest of times. His never-ending sense of humor and stories will always be remembered.

*The wealthiest person is a pauper at times
Compared to the man with a satisfied mind*

~ Porter Wagoner, 1927-2007

CONTENTS

	Introduction	ix
1	Becks Creek to Cripple Creek	1
2	On the Move Again	65
3	California	89
4	Aunt Minnie's	143
5	Up In the Air and Under the Sea	179
6	Mediterranean Bound	223
7	Discharged	247
8	A Growing Family	285
9	On the Road With Jeanie	337
10	A Good Life	357

INTRODUCTION

There's something to be said for those of us who have spent a big part of our lives living in poverty. Most people of the generation that followed ours have no idea what it's all about. Living in a country as prosperous as ours, it's hard for most of them to imagine that some of us had to live as poorly as we did.

You just don't get the real picture unless you've been there. Unless you have experienced learning to "make do" with what little you have or have had to struggle just to have the basics, like a roof over your head, decent clothes to wear, and food on the table.

Many have never had to wear *hand-me-down* clothes, clothes that didn't fit, or patched clothes to school. Or wear worn out shoes with cardboard in-soles to keep out the dirt. Or face the embarrassment of wearing run-over, unpolished shoes with soles held on with bailing wire and socks pulled down and tucked under the toes to hide the hole in the heel.

Most people living in these times have never experienced doing their homework by the light of a smoky kerosene lamp, or taking sandwiches made with biscuits and salt pork in their school lunches, or even wearing their sister's shoes like I did because my folks didn't have the money to buy me a new pair. Most have never had to carry water from a stream or cistern for family cooking and bathing, or experienced the kids in the family having to sleep in one bed to keep warm, not having enough mattresses and bedding for each one to have his own bed, and sometimes sleeping with a pre-heated old flat iron to keep our feet warm. This is a story of my family, a story of how we had to live during and after the Depression that followed the financial market crash of 1929. This is about the hardships we faced and how we had to cope or "make do" in situations when we were struggling to have the basics as we moved from state to state as Dad took different jobs trying to make enough money to get us a decent place to live. I have tried to tell it like it was, going back as far as I can remember.

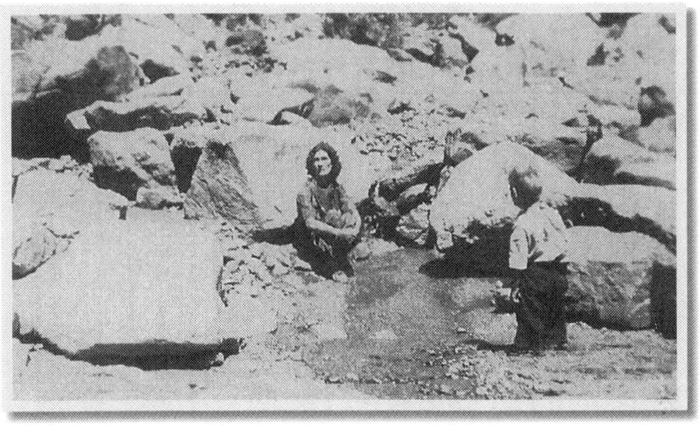

1 BECKS CREEK TO CRIPPLE CREEK

In the summer of 1986, my sisters Laura and Lila and my mother and I visited Williamsburg, a little mountain town eighteen miles north of Jellico, Tennessee that sits along the Cumberland River off of Interstate Highway 75. It was a real enlightening trip for the four of us. We hadn't been in Williamsburg since the 1930s when I was about four years old.

I was born in a small cabin near Becks Creek on Becks Creek Mountain about ten miles from downtown Williamsburg, Kentucky. The cabin stood on the forty-acre homestead that belonged to my grandfather Will Jones, the man who was known in Williamsburg and all over Whitley County and Jellico, Tennessee as "Bumper Jones."

There were seven children in our family, four boys and three girls. The boys, starting with the oldest, were Bill, David, Jim, and Tom. The three girls, starting with the two oldest who were identical twins, were named Leta and Laura, and Lila was the youngest.

Williamsburg is noted for its fine college, the "Cumberland College," considered to be one of the best in the nation. It has two one-way bridges spanning the Cumberland River, one for entering town and one for leaving. Old brick buildings line the streets, and the picturesque courthouse with its park-like setting is not too far from the river. The college too has all brick buildings, both classrooms and dorms. The rolling hills and lush green lawns of the campus stand out as you head toward Becks Creek Mountain.

Most of Dad and Mother's families grew up in or near Williamsburg, and there are still many Creekmore and Jones families in and around the area. The Creekmore name is on just about every business, from markets and shops to produce trucks. There is a large Creekmore cemetery about ten miles from town, not far from where the town Bon Jellico, a coal-mining town, used to be. The little white Baptist Church that stands near the cemetery is where my parents were married in 1921.

Both Mother and Dad were born in Williamsburg, Dad, on the Fourth of July and Mother, on the fourteenth of June, Flag Day. You might say they were real patriots with their birth dates and all. Mother al-ways presented her big flag on those dates. Though my mother was part Cherokee, we were all taught to respect the flag and stand up for our country. All of us boys who were able joined the military service when we came of age.

Both the Jones and the Creekmore cemeteries are on the same road on Becks Creek Mountain, only a couple miles apart.

The Jones cemetery, much smaller than that of the Creekmore family, is split by the road that winds around the mountain. The moss-covered tombstones of my grandmother Permeile Jones, my grandfather Will "Bumper" Jones, and their only son John can be seen from the road.

Some of the tombstones were just small pieces of sandstone with initials scratched in them—no words, no dates, just initials. When the four of us located the graves of her parents on our trip, my mother stood between them crying.

I remembered very little about Williamsburg, but my mother and sisters remembered plenty—the red brick buildings in the old part of town and the shops and stores downtown. Mother showed us a little brick house not far from the post office where her family lived when her dad remarried not long after the death of her mother.

We decided to park and have some lunch and then walk around town so Mother and the girls could reminisce a little as we looked at the some of the old brick buildings that Mother and her family had frequented when she was a young girl. We parked in front of the Court House and were walking across the lawn in back of it toward a little restaurant on the street behind it when we noticed two middle-aged men in bib-overalls sitting on a bench talking. The grass in front of them was covered with scraps of wax paper and wood shavings. As we approached the two, we noticed that they were whittling and spitting snuff as they talked. We said hello as we walked by and the one on the right spit out a big spurt of snuff, cleared his throat, and in a loud voice said to Mother, "Don't I know you?"

She stopped and looked at him, studying his face, trying to remember who he might be.

"Ain't you Bumper Jones' daughter?" he asked.

We couldn't believe what we were hearing. Mother said she couldn't remember him and asked him how he knew that her dad was Bumper Jones.

He said that he was sitting in a buckboard (a horse-drawn carriage) the night the shootings took place. He said that he was with his family at the Junior Mechanics Celebration in Corbin when my grandfather was shot by a drunken relative named Ed Moss Jones.

His death, a real tragedy, was brought about by a fight between his son John and a second cousin, Ed Moss Jones, during a Junior Mechanic Celebration. John and Ed had been drinking moonshine and arguing when Ed pulled a .38 pistol and began firing at John, hitting him in the leg. Then, firing wildly, he shot and killed his own brother Jimmie standing near John. Grandpa Jones had gotten between John and Ed and grabbed Ed's hand to keep him from killing John. In the struggle, he was shot in the stomach and died in agonizing pain the next day. He was thirty-five years old.

Needless to say, the family was devastated. There were several newspaper articles that came out in the Corbin Times Tribune on October 3, 1924. Here's an excerpt from a couple of them:

> "Part of next week will be given over to misdemeanor cases after which the felony cases will come up for trial. Twenty-five misdemeanors are slated for Monday, twenty for Wednesday and twenty for Thursday. A special judge will be secured in two cases, which will come up on the thirty-first day of the term, Judge Tye disqualifying. The trials are that of

Ed Moss Jones for the killing of Will Jones and his own brother Jimmy and for Jeff Cox for the killing of Grover Sharp.

Corbin Times Tribune, November 7, 1924:

"...Ed Moss Jones was given a four-year sentence this week for the killing of Will Jones on Becks Creek a year ago at a Junior Mechanic Celebration. It will be remembered that Jimmie Jones, brother of Ed Moss Jones was also killed in this same difficulty and that Johnny Jones was also shot. The actual shootings took place in 1923 a year earlier."

After Mother had talked awhile with the man who remembered her, Laura and Lila walked on each side of her toward the restaurant, attempting to hold onto her arms as she stepped off the curb. She said, "Let go of me, they'll think I'm old," Laura looked at Lila and smiled as we started into the restaurant. We really enjoyed our lunch that day.

We ordered hamburgers for Laura, Lila and me, a hotdog for Mother, and a milkshake for each one of us. The waitress brought the three hamburgers on buns and set all the fixin's on the table and we made our own sandwiches.

They turned out to be some of the best hamburgers we'd had in years and the big milkshakes we enjoyed were like we like those served in the ice cream shops in the 1950s. They were made in a big Hamilton Beach mixer; our glasses were filled and the remaining ice cream was set before us in the tall stainless steel mixing containers.

We walked around town from one end to the other, stopping at some of the gift shops to look for things for my two sons, Jim and Gary. We checked out the little red brick house that was on one of the streets behind the courthouse–the house where Mother's family had lived after her dad re-married. His second wife's name was Miss Laura Steele, "Ma Jones." She gave birth to Mother's stepsisters Mildred and Willie. They all addressed her as Ma Jones because the youngest twin in our family born in 1925 was also named Laura.

As we walked out toward the college and Becks Creek Mountain, we stopped near the gate of one of the big mansions that was set back off the street. As Laura and I walked farther down the street to look for some more shops, Mother and Lila struck up a conversation with the caretaker that stood near the gate. When we came back, Mother had us take their picture and then all of us headed toward Becks Creek Mountain.

♪

Mother often talked of her life growing up in and around Williamsburg; she talked mostly of her immediate family and she often talked about her dad, Bumper Jones as he was called, and her little mother Permeile who had a serious heart condition. Mother told us that her dad would often take Permeile in his arms and carry her when they walked any distance or walked up a steep hill and she became short of breath. He would carry her until he came to a fairly level or downhill stretch, and then they would all rest. She told us what wonderful parents she had and how special her dad was to all the family.

He worked in the Bon Jellico coal mine when the mine was going strong in the early 1900s. Her wonderful little mother passed away when Mother was in her early teens. She told us about her brother John, the youngest in the family who would go to dances, drink too much moonshine and get into fights, and come home all skinned up with cuts and bruises all over his face and arms.

Will (Bumper), his wife Permeile, two daughters, Lillian and Minerva (Minnie), and his son John, the youngest in the family, lived in one of the miners' shacks in Bon Jellico, a coal mining town that is no longer in existence. The coal mine had opened in 1912 and closed down in 1937. She said the shacks were small and had no running water or electricity and had small two-holer outhouses with no lights or running water.

Her dad and all of the mine workers were paid in script¬–little metal tokens that were only to be spent at the Bon Jellico Mining Company store, like the old song that was popular in the 1950s, "Sixteen Tons" sung by Tennessee Ernie Ford with the lyrics, "I owe my soul to the company store." Her dad only made a little over two dollars a day. He was a religious man who seldom missed taking the family to the First Baptist Church on Sundays.

In the evenings, he would sit for hours on the front porch playing old folk tunes on his five-stringed banjo. The names of some of the tunes she remembered him playing were "Sourwood Mountain," "Cripple Creek," "Rubin," "Greencorn," "Keep Your Skillet Good and Greasy," and "The Little Whitewash Cabin," just to mention a few.

I suppose that's one of the reasons I've always loved the sound of bluegrass music and that special ring of a five-string banjo. I guess you could say it's in my blood.

Mother knew how well I liked the sound of a banjo and bought me a nice five-string for my fifteenth birthday in 1949. (I still have the banjo, and still can't play it like it should be played.)

She said her dad would sit on the little front porch, comfortably leaning back in his chair with his eyes closed, holding his banjo, picking song after song, with one foot resting on the little skinny porch rail, the other tapping time with the music.

He had a fine pair of mules he used to work the land and a nice horse and buggy for the family transportation. Although he was a kind man, he was noted for his quick temper; when he got mad about things beyond his control, he would bump his head against the wall instead of cursing, which is why most of the town people in and around Williamsburg and Corbin knew him as "Bumper" Jones.

Permeile died when Mother was still in her teens. Bumper remarried and became the father two more children, my mother's half-sisters, Mildred and Willie. Willie was born after his death.

♪

Mother had relatives in the Civil War, in both Confederate and Union armies. She told of how two of her aunts had dealt with Union soldiers trying to break into their cabin when they were alone. They were just getting ready to do a wash when it happened. They had their scrub board, lye soap and the lye they used in their water all laid out on the cabin floor. They heard footsteps on the porch⌐, and someone trying to open the door that they had barred from the inside; one of them quietly climbed the stairs to the bedroom in the loft and peeked down, seeing the Union soldiers on the porch.

The soldiers began to yell for them to open the door. By this time, the two women had added lye to the boiling water on the stove as the banging and yelling grew louder and louder. They quickly but quietly hoisted the boiling lye water to the loft. By then, the soldiers were pounding on the door, trying to break it down with their rifle butts; as one of the women held the shutter open, the other dashed the boiling lye water on the soldiers below, sending them screaming and running back into the woods.

My dad, William Ezra Creekmore, never told us anything about his family; what little we were told came from Mother and the research we did in later years.

I was told his mother died of old age and his father died of galloping consumption, a lung disease that we now call tuberculosis. Mother and Dad dated for a couple of years before they married. Dad was a good mechanic and machinist and did fine lathe work. He really liked guns and was quite a marksman with both handguns and rifles. Mother said when they went to competition shoots or turkey shoots, he would nearly always take first prize. He was such a good shot that she at one time let him literally shoot rings around her feet with his luger. Before they were married, he had spent some months in the army; he wasn't in long till the Armistice was signed and he got discharged.

A short time later, mother's sister Minnie married Grant Parker Carroll, a close friend of Dad's, and they moved to Cripple Creek, Colorado. It wasn't long be-fore Mother and Dad decided to follow. Aunt Minnie wrote mother telling her of jobs that were available in and around Cripple Creek. Uncle Grant was working for Parcel Post and he said he would help us out until Dad found work. This was the Depression era and jobs were scarce.

When the family moved to Cripple Creek, there were four children: Bill, the oldest, twins Leta and Laura, and the youngest of the girls, Lila. David, the fifth child, was born in Cripple Creek in December 1932, a couple years before I came along in May of 1934. Mother and Dad ran a restaurant for a short time in Cripple Creek before going back to Williamsburg when they couldn't make a go of it.

Then, Dad heard they were hiring mechanics to work on the assembly line at the General Motors plant in Detroit. His application was accepted and he moved the family to Detroit. He began making a pretty good wage; they were able to rent a nice house and he had plenty of money to support the family. He sometimes drove a new Cadillac, Olds, Buick or Chevrolet home for a road test. They were, as they used to say, "livin' pretty high on the hog." Everything was going well until he came down with polio.

Polio had become epidemic in and around Detroit at the time. He had just sent the family back to Williamsburg on vacation when it happened. He woke up one morning with a violent headache, and as he was shaving, he was having trouble standing; he thought it was because of the severe headache, and he thought it would go away as soon as he had some coffee. He managed to drive to work; but, by the time he got up on the assembly line, he was having a tough time standing up. After he fell down a couple of times, his supervisor had him taken to the hospital in Detroit and after a thorough examination; he was diagnosed as having polio. By then he was losing the use of both his arms and legs.

When Mother got word that he was so sick, she took the first train she could get and brought him back to

Williamsburg where she could take care of him. The doctors in Detroit said he most likely would be an invalid for the rest of his life and would never walk or use his arms again. He was home about two weeks, when she got more devastating news; she started getting letters addressed to him from some woman in Detroit she had never heard of. He told her that the letters were not important, and that she should just throw them in the trash. Instead, she opened them to find that he had been dating this woman and had promised her he would divorce my mother and marry her.

All those times, including Saturdays that he was supposed to be testing cars or working on them, he was dating his new love. He had sent the family on a so-called vacation so he could be with her. When Mother read the letters to him, he begged her to forgive him. There she was with a crippled and maybe even dying husband, five little children, and no insurance or income.

When my dad first came back to Becks Creek to be cared for, she would walk ten miles to Williamsburg, to wash and mend clothes at the jail—or whatever needed to be done—and then she would often walk home carrying groceries. She would work with him massaging, straightening, and exercising his arms and legs. She also managed to get a therapist and a chiropractor for a few visits, but she managed most of it herself. After months of painstaking care, he was able to stand and take short walks using crutches.

Shortly after that, the family moved from the homestead on Becks Creek to Savoy, a little town not far from Wil-liamsburg where there was a bit more opportunity. She worked at a restaurant and took in washings and ironed clothes to support the family. Why Dad's fami-ly didn't help us out, I'll never know. In spite of all his wrongdoing, she was determined to get him walking again. I guess she was sticking to her marriage vows—for better or for worse, till death do us part—she had been raised a Christian, and that helped her through all the heartaches and hardships that she faced. She didn't attend church often after they were married, but she stayed awake late and read the Bible just about every night.

I didn't know until years later that Mother had only finished the third grade. She taught herself to read and write. During World War II, she kept track of the war by reading the latest newspapers, listening to the radio, and writing letters to aunt Minnie.

One afternoon, the family was out walking and we had our dog Nig with us. The dog ran toward Mother, brushing Dad, causing him to lose his balance and fall before Mother could get to him. When he was struggling to get back up, he let out a painful scream; the tendons in his legs that had been drawn up for so many months had suddenly straightened out. As he wobbled to his feet, Mother ran over to him, and physically supported him until they got back to the house. After that incident, he was able to walk much better and soon walked without crutches, using only a cane.

Poverty Gulch

One day a letter came from Aunt Minnie in Cripple Creek. She said Uncle Grant could get Dad a job if we were there. Soon we were on our way back to the high altitude town in the Rockies. Dad had managed to get a car, a 1928 two-door Chevy sedan, and he was able to use his legs, though he had hardly any strength in his arms. He had no trouble steering, but he did have trouble raising his left arm to put out the window for his turn signals.

We moved back to Cripple Creek in 1938 just before my fourth birthday. We were told that Cripple Creek, a boomtown during the gold rush days, was named after the stream that ran through it. A rancher had named the stream after one of his cattle, while cross-ing to the other side of the stream, had fallen and crippled itself when walking over the slick rounded rocks.

Cripple Creek had once been a good-sized city. In the early nineteen hundreds, there were said to have been over 55,000 residents in the Cripple Creek District. When we moved there in 1938, there were lots of vacant old brick buildings, but there were still hotels, restaurants, filling stations, grocery stores, bars, and banks.

Winter in Cripple Creek was quite an experience. We had seen cold weather in Kentucky, but nothing like the freezing weather in Cripple Creek. I'm sure the cold winters in Cripple Creek, Kentucky were never like those in Cripple Creek, Colorado. When we first arrived, we didn't have the money to rent a decent house, even with uncle Grant's help.

Our first house was in Poverty Gulch, a name quite fitting for our situation, a part of town where folks like us were living in rent-free tarpaper shacks. The shacks were the unfinished cabins that had been started to house some of the miners before the Depression. Some had tin roofs and doors, but no glass in the windows, and pieces of tin covered the small window openings. The shacks had old newspapers for insulation and newspapers pasted together in layers taking the place of sheetrock. They had no siding and the only protection from the weather was the heavy tarpaper tacked to the outer surface of the two-by-four studs that framed the little shacks. The cabins had no running water or inside toilets. We were lucky to get one with that was within walking distance of town.

A rancher by the name of Bob Womack had discov-ered gold in a Poverty Gulch in 1890. When the last of the mines closed down in 1961, over $500,000,000 in gold had been dug out of the hills in and around Cripple Creek.

The historic gold-mining town was built on the rolling hills not far from Pikes Peak. Most of the streets were on hill slopes, and you could roll a bottle from one side of the slanted streets to the other.

When we went to town from our tarpaper shack, we had to cross over a train trestle. The trestle bridged two rock faces that had been blasted out so the train could come right up between them into town. One Saturday evening as the sun was going down, the twins took me downtown to see my first movie, Snow White and the Seven Dwarves. The popular kids' movie that was produced in 1937 had been showing in theaters across the country some months before we moved to Cripple Creek. My sisters had planned on taking Dave, Lila and me to see it, but Dave and Lila were at Aunt Minnie's when Laura and Leta won three show tickets – the first place prize in a singing contest held that Saturday afternoon on stage at the theater. They were both very talented singers and dancers. The songs they sang were sung in perfect harmony and they came away with our show fare and had scraped up enough money for popcorn and ice cream.

I had no idea what I was about to experience in that walk to the theater that evening. As we walked down the path toward town, we could hear the big steam engine as it puffed its way into town, getting closer and closer to the trestle.

Just as we had reached the middle of the trestle, the big steam engine, pulling I don't know how many cars, passed not ten feet below us causing the trestle, the ground, and everything around it to tremble from the weight of the big locomotive. The hot smoke and scorching steam and sparks blew up between the planks from the train's smoke stack, along with the sound of clanging bells, which made for quite a thrilling (or you might say scary) experience for me. I was holding on to the twins for dear life.

In the end, the trestle scare was really worth it; I got to see a real good movie and I had plenty of popcorn and enjoyed my first ice cream cone, a big deep cone full of strawberry ice cream before we started home. On our way back, I didn't mind walking over the scary trestle at all. We talked about the movie all the way home.

Box Canyon

We didn't live in Cripple Creek very long before we moved to Box Canyon. Dad was doing security work in town, keeping an eye on certain buildings and such; and he and Uncle Grant shared my uncle's old Star pickup. Uncle Grant, or "Uncle Parker" as we called him in those early years, was real kind to our family, but quite overbearing with his own. He was rough on the kids. He was constantly quarreling with Aunt Minnie, yelling at her and arguing with her. They swore at each other when he ordered her around. He had bought some land in the canyon and built a small cabin on it that was not too far from an old bar and stage coach stop, which sat not far from the creek at the bottom of a hill.

We moved into the old stagecoach stop, paying very little rent; it had been a bar and small hotel. It was pretty run down to say the least. It had only one window that still had glass; the rest had been broken out and we covered the openings with scraps of tin to keep out the wind and snow. We slept in the basement because it had very few leaks and was the warmest, driest part of the old coach house. Snow and sleet blew in through the cracks around the windows, but very little of it dripped into the basement. It was so cold, we slept in our clothes to stay warm; we had no pajamas or winter long underwear. We didn't have mattresses and quilts for us to have separate beds, so the boys slept on one mattress; and Lila, the twins and me (being the youngest) slept on the other. Every night the twins would sing me to sleep with songs I've never forgotten like "My Old Kentucky Home," "When It's Spring Time in the Rockies," "My Blue Ridge Mountain Home," and the one song that they always sang in perfect harmony called "My Old Cottage Home." They would also sing "White Cliffs of Dover" because the song mentioned my name ("and Jimmy will go to sleep in his own little room again").

Mother and Dad had their own small room. At night, when the icy winds blew down through the canyon, the old house came alive with all kinds of strange noises. There were scary noises coming from the balcony, a part of the house that was off limits to us. The loose boards banging and low moaning and screeching noises could be heard as the blasts of wind hit the boards and the tin covering the windows. All of this, along with brother Bill's ghost stories and the noise of rats dragging things in the attic and room above us made the old house pretty scary after the kerosene lamps were snuffed out.

One of the roughest parts living there was when we had to use the old outhouse, an outside toilet, especially at night when it was snowing or sleeting. The outhouse was about thirty feet from the main house. We would put on a coat and shoes, get Dad's flashlight and head out. Needless to say, we pretty much did our business before dark. We always felt safe because Dad always had his pistol, his loaded German Luger, at the head of his bed.

Uncle Grant's little ranch consisted of a small cabin he had cobbled together on the side of a hill in Box Canyon using freight pallets and scrap lumber. He had also built a nice chicken house with part of it serving as a turkey roost for a pair of turkeys, and he had added on goat and pigpens. The animals and the birds were fed garbage from the hotels, restaurants and grocery stores in Cripple Creek. He would drive down the alleys and unload their garbage cans into two empty oil drums in the back of his pick-up. He separated the wet stuff or "slop" from the breads and vegetables, and after filling the hog troughs, he would then scatter the bread and vegetables for the turkeys and chickens, and throw some to the goats.

Not being of school age, I was able to go with him on his garbage runs. Those trips were fun trips for me. Before we started back to Box Canyon, he would always stop at a grocery store and buy me a candy bar or cookies. He would whistle or sing the whole trip, or recite some of his funny poems for me. One of his poems was called "Jim Bleacher, the Sunday School Teacher." As I remember, it goes like this:

> *Jim Bleacher, Jim Bleacher, a Sunday school teacher; his dad was a minister, too*
>
> *He took a notion he'd cross the ocean and teach to the cannibals of Tim Bucky Two.*
>
> *The cannibals took him, was going to cook him when Jimmy broke loose from the bunch,*
>
> *They all heard him cryin' as Jimmy went flyin', "Sorry boys, can't stay for lunch."*

Uncle Grant was a little man about five feet tall, but very strong for his size and quick at getting things done. We liked going to our Aunt and Uncle's place and playing with our cousins and enjoying the animals. They had silly goats that could almost jump straight up in the air, and the kids had given them names; how and where they came up with their names is beyond me. Teak and Gunateak, Brently, Babe, and Fart Damn Stinker were all names chosen by Nancy, Dave and Grant Junior.

The goats had the run of the hills when they were let out to graze during the day, and they were rounded up before dark to be fed and milked. We enjoyed eating at their house. Uncle Grant had steady work and they had more of a variety of food than we were used to. Aunt Minnie would fry up a big platter of hamburgers and have plenty of biscuits and mashed potatoes and gravy. We were used to having biscuits, mashed potatoes and gravy, but not those delicious hamburgers followed with an ice-cold glass of goat's milk and a big slice of her special dessert, banana cream pie. We loved the banana cream pies, but drinking goat's milk was another matter; we had to force ourselves to drink it. Unless you've been raised on it, it takes some getting used to with it is strong smell and taste.

Our families were very close, more like brothers and sisters than cousins. I think we really liked being around Uncle Grant more than Dad. Dad certainly wasn't very affectionate. He would walk up behind me when I was playing, and pull the hair on the back of my neck making me almost cry. Lots of times instead of calling me Jimmy when he told me to do something he called me Banjer Eyes (meaning banjo eyes) because of my big, round eyes.

The older kids had fun playing in the creek or riding their cobbled-together coasters down the steep hill toward the creek. The coasters were made from scrap lumber taken from behind the chicken pen and some buggy wheels cousin Grant had brought home from the dump. Back then everybody that lived in the country had his or her own dump, usually a low place in the ground not far from the house. Sometimes when they raced the coasters, they had some pretty bad spills and Aunt Minnie and mother would apply the turpentine and sugar to the wounds and wrap them in clean rags, usually torn from worn-out sheets. The coasters had mismatched buggy wheels, no steering wheels, and the front axles were nailed to a two-by-four with a pin in the center and a rope nailed to each end. When they sat on the coaster, their feet rested on each side of the board and the steering was done with their feet and the rope. It was like holding onto the reins of a galloping horse, making it pretty difficult to steer clear of some of the big rocks on the hillside.

In the summer about once a week, depending on how hot the weather was, we would take my cousin's wagon, a beat-up old Radio Flyer, an ice pick, and a flashlight and head for the "ice" cave. We had to go down the hill and cross the creek to a cave that went back into the side of the mountain.

The part of the mountain where you entered the cave was shaded on both sides receiving very little sunshine both summer and winter, so there was very little melting when the spring thaw came. When my eyes adjusted to the semi-darkness and Nancy held the flashlight, pointing the beam to the back of the cave, I could see the solid wall of ice about forty feet back. Nancy held the flashlight as Grant chipped away at the ice, chipping as many big chunks as possible to fill the old wooden ice box that Aunt Minnie had in her little kitchen.

Obviously, we kept the icebox full without complaint, since going to the cave was an adventure for us. We were never allowed to enter any of the other caves that were in Box Canyon. There were too many abandoned mine shafts in the hills around Cripple Creek. Not knowing, a person could easily fall hundreds of feet to their death down one of those shafts. When the kids went out to round up the goats, they were always warned of these dangers.

Springtime was a beautiful time in Box Canyon. As the snow melted, there would be patches of wild flowers popping up around the huge boulders and outcroppings of rocks throughout the canyon. The cool mountain air was scented with the smell of pines and wild flowers. I'll never forget the rumble of the creek from the spring runoff. By mid-spring, when the heavy winter snows and large chunks of ice began to melt and break up, along with rocks and even small boulders rolling down into the creek bed, a special roar was created that could be heard way back up on the mountain. It definitely wasn't a time to try to cross the creek even for those with experience.

One hot day when the twins were enjoying a walk in the canyon, they noticed a cave going back into the hillside where the goats had been grazing. They decided to explore a little, knowing that they would get in big trouble if Mother or Dad found out. The first few feet they could see; but, after about twenty feet, it was just black and they couldn't see a thing. Instead of turning around and coming back out, they decided to go just a little farther. They started tossing rocks ahead of them to see if they were nearing the back of the cave. They did this a few times, listening for the rocks to hit before going any farther. After about the third move, they pitched rocks ahead to check for solid ground. A half a minute later, they heard a splash as their rocks hit the water in the bottom of the open mineshaft. This scared them, and they turned and hurried back toward the entrance.

A week or so later they gathered enough courage to tell the family about their close call as we all sat at the supper table. Had they not been cautious, the walk in the mine could have cost them their lives. They were lucky too that they didn't get a good switching from Mother for doing what they were told not to do. But Mother knew that they had learned their lesson by having their scary close-call experience.

After the twins told us how close they came to falling to their death, all of kids were reminded of what we had been warned about, and we never tried to explore the tunnels. The abandoned mines in Arizona and New Mexico held people as well as animals that had fallen hundreds of feet to their deaths in open mine shafts, not knowing they were there. Many of the shafts in those states are covered over with rotted planks in the shade of mesquite trees and have no warning signs.

♪

In Box Canyon there were all kinds of game. On some of their hikes, Bill and the twins had seen not only deer, but also mountain lions and bears as well in the rim rocks and near the stream. Though compromised by the polio, Dad was able to walk on some of the deer trails that ran along the creek in the canyon where he and Bill would go hunting, and they brought home plenty of deer meat for the two families. The ranchers in the area knew that Dad and Bill were illegally hunting the deer out of season, but they never turned them in or said anything about it. I guess they realized how poor we were and knew that the fresh deer meat was used to feed our families.

When Mother and Dad took walks in the canyon, the twins and Bill would watch after Lila, Dave and me. I remember spending most of my time rocking back and forth in our beat up old rocking chair singing songs like "Birmingham Jail" and "Hallelujah, I'm a Bum."

There were some deep holes washed out around some of the big boulders from the spring runoff. The water was deep enough for swimming, if you could stand to swim in the icy water. All of our water for drinking, bathing, and washing clothes came from the creek. The streams were not polluted then, and very few people got sick from drinking the water from the mountain streams.

Colorado, noted for its aspen trees, definitely didn't exclude Box Canyon. Nearly everywhere there was a mountain spring, there were aspens growing; they dot-ted the hillsides and grew near the base of huge boul-ders where the springs bubbled out. The San Isabel National Forest, south of the town of Florence, was mostly made up of aspen trees.

Living near my aunt and uncle and their family was a real help for us; we look back on it all with much appreciation. They not only helped us financially, but also helped us have a positive outlook on our situation, which meant a lot. We never let our minds dwell on being poor, we were always thinking of fun things to do. As kids, we never thought much about the serious situation we were in.

Uncle Grant, like a lot of people back then, re-soled the family shoes. The shoes and boots he repaired were re-fitted with new Cat's Paw soles and heels, a brand still in use today. He had a regular cobbler's last, a device that held the shoes as he worked on them. It was like an upside-down iron foot that the shoe or boot slipped over and it was held in place while he made the necessary repairs. He had all the tools for the job. He would remove the worn out soles and heels, take his special rasp (a cobbler's tool) and sandpaper and rough up the area that was being re-soled. With shoe or boot still on the last, he would then glue, clamp and tack down the new soles and heels. After leaving them overnight to dry, he would trim off and sand down any rough edges and polish the shoes. He would usually be whistling or singing as he worked. The finished product turned out as good work done at any professional cobbler shop.

Shoes were an expensive item and it was rough on my brothers and sisters not having decent shoes to wear to school. We had barely enough money to live on and being able to buy suitable shoes for five school kids was really tough. Before Uncle Grant started repairing the family shoes, Dave had started school with shoes so worn out that after the first few days of school, his teacher couldn't stand to see the little boy coming to school practically barefoot with cardboard inner soles and what was left of the shoe soles wired on with bailing wire. Without saying a word to Mother, she bought him a new pair. Mother was in tears when he came home from school wearing new shoes. She was embarrassed but very grateful. She let the teacher know right away how much she appreciated those shoes.

Penrose

We didn't stay long in Box Canyon. One winter was about all we could take and we had to find someplace where Dad could find a better paying job. We had to move where the winters weren't so bad. We needed to move to some place where we could walk to the store and a place where schools were closer.

Dad heard that they were hiring mechanics and machinists in the smelter town of Pueblo near the New Mexico border. What we called "smelters" were places where they refined ore. The folks had made up their mind to get away from the freezing place in Box Canyon. We loaded up our old 1928 two-door Chevy sedan, the same one we had come to Colorado in from Kentucky, and we headed south to the little town of Penrose.

Penrose was what we called a "wide spot in the road" about forty miles south of Cripple Creek. We took Phantom Canyon road out of Victor, which took us south to Florence, and from there we went east to Penrose. Phantom Canyon was used a lot in the early days of the gold rush and in the early 1900s. Several trains steamed up through the canyon from Florence to Victor through Elkton and Anaconda to Cripple Creek, about a 5000-foot climb over forty miles or more of steep grades. Victor had once been the home of the famous journalist Lowell Thomas. Back in the days of gold rush, a train known as the Gold Belt Line ran from Florence to Cripple Creek, carrying passenger trains on a daily basis, three cars each way.

When we got to Penrose, the job that Dad thought he had in Pueblo didn't work out, so we were just about as hard up as before. He finally found a few security jobs that hardly paid anything, and Bill, the oldest, worked the summer carrying water to the workers employed by the Works Progress Administration (WPA), a central part of President Franklin D. Roosevelt's "New Deal," enacted by Congress in 1935. WPA not only worked on roads and bridges, but also built outhouses. We were eligible for "relief," as welfare was called in those days, but both parents were too proud and embarrassed to apply for it. Finally, out of desperation, Mother had the twins fill out the application for it. We were then able to re-ceive staples like dry beans, flour, powdered milk, baking powder, lard and salt, all of which helped us a lot.

The little two-bedroom house in Penrose sat next to our landlord's cow pasture. We all remember the day when we didn't have salt for the beans, so Bill climbed over the fence with a hammer and broke chunks of salt from one of the pale green salt blocks that the rancher had set up for his cattle. He was lucky the herd was at the other end of the pasture and he didn't have to worry about being gored by a charging bull. He brought back enough salt to last several days. Getting the basics we needed through the relief program prevented more stunts like that. We weren't there too long before Uncle Grant, knowing that Mother was pregnant with another child (Tom), brought down one of his milk goats so that the new baby and our family would have fresh milk. Because most of us disliked goat's milk, we were relieved when in the middle of April, a little over a month after Tom was born; Uncle Grant came down for a short visit and took the goat back to Box Canyon.

The Ride Up Phantom Canyon Road

One weekend, Uncle Grant brought the whole Carroll family for a visit. They came down from Box Canyon in a flatbed truck with sideboards, a tarp stretched over them from the cab to the tailgate. The kids had ridden down Phantom Canyon Road sitting up against the cab on pillows in the back.

After their visit, Mother agreed to let Dave and I go back home with them. That trip back up Phantom Canyon Road was one we would never forget.

When we left Penrose late that evening, we could see lightning flashing way back up on the mountain, which was a pretty good sign it was getting ready to rain in the higher elevations.

By the time we all got loaded in and said our goodbyes it was almost dark. After Uncle checked the tie-downs on the tarp and made sure that all of us kids were settled on our cushions, including their dog Big Paws, we were given our instructions not to be quarreling or fighting or trying to stand while the truck was in motion.

He fired up the old truck and we headed for Phantom Canyon Road. I don't remember the make or model of the truck, but it wasn't the old Star with the open cab that he and Dad had shared when we lived in Box Canyon.

By the time we got started up the mountain, it had begun to rain. The going was pretty slow since there were no guardrails on the gravel road, just turnouts where the road was too narrow for two vehicles to pass each other. The road had been blasted out around the side of the mountain, a shelf road. We more or less had to travel in the center of the road and stay off the shoulder. In some stretches, if you got too far over, you were in danger of falling hundreds of feet to the stream below.

The thunder rolled and the lightning flashed, lighting up the whole side of the mountain clear down to the stream at the bottom of the canyon as we all bunched up against the cab, pulling our coats up tight to keep away the rain that was blowing under the tarp in between the sideboards.

As the lightning flashed and the claps of thunder echoed off the canyon walls, we slowed down and pulled in close to the bank and stopped; we peeked through the back window in the cab and could see the lights of another car coming around the curve ahead. As the big lights got closer, we could see that it was a good-sized truck loaded with lumber. We could see too that the road at that point was too narrow for both trucks to pass one another.

Since we were the uphill traffic, we had to give way to the downhill traffic, giving the loaded truck the right of way. As the big truck dimmed its lights, Uncle Grant slowly started backing down the hill to find a turnout. Just before we got to the turnout, he stopped again and waited for the other truck to get close enough so that he could talk to the driver. After they talked, we again started to back into the turnout, which had no guardrail and was on the outside track facing the canyon. We let the loaded truck have the inside track against the mountain. They had decided that with the weight of the bigger it truck, the outside shoulder might give way and cause it to slide off, sending it to the bottom of the canyon. As the lightning continued to flash, we were all pretty scared when we looked out and could see the sheer drop-off to the streambed several hundred feet below. Like I said, it was a truck ride we would always remember.

♪

On March 1, 1939, Mother gave birth to my little brother Tom. Thomas Ray, the second one in the Creekmore family to be born in Colorado. We were still struggling to get by, and the twins were now working to help with the family's income. They had to start school later than many others because they had no decent shoes to wear.

They worked with mother picking cherries while Lila took care of Tom, Dave and me. We would sit playing on a blanket in the shade of a cherry tree not far away from the orchards. Dave, under the supervision of the twins, would help place the picked cherries in boxes.

One day Lila stayed home with Tom, Dave, and me. She decided to put on a pot of beans and have them ready in time for supper. She had never before cooked without the supervision of the twins or mother, and she never had cooked a pot of beans. She misjudged the size of the pot from the very start. She started out with what must have been five pounds of beans. She noticed as the water began to boil that they were swelling up and running over the sides of the pot. By the time she managed to get them all cooked, she had every kettle in the kitchen full of beans. The twins and Mother couldn't keep from laughing when they came home to supper and saw every pot full of pinto beans. Needless to say, they didn't have to cook beans again for several days.

Our meals were usually made up of beans and biscuits or cornbread with wild mustard or turnip greens, and sometimes, wild asparagus. We often had fried salt pork and sometimes fried chicken or chicken and dumplings. We also took fried salt pork and biscuit sandwiches in our lunches when we attended school. Mother would slice the salt pork like bacon and then place it in a skillet of boiling water to boil most of the salt out of it, and then fry it like bacon. When we all scooted up to the table with our mismatched wobble-legged chairs, we were quiet and mannerly. Mother was very strict about our table manners. There was no horseplay allowed. No elbows on the table or giggling or cutting up. As the platters of food were passed, no one took a large portion. And. we never took the last portion or last piece of chicken or meat off the platter without asking first if any other family member wanted to share it.

Our frugal suppers were usually eaten by the smoky light of a kerosene lamp. For dessert, Mother made pies, mostly cobbler pies. When she had no fruit for her cobbler pies, she made fried pies. These pies were made with vinegar and sugar and cinnamon.

She also baked large loaves of light bread and buns made with lots of yeast that we enjoyed with our pinto beans. This was all done on or in a wood-burning stove. Some of the cook stoves like one I remember had water reservoirs on the side for heating water.

Our baths were usually taken in the kitchen; the bath water was carried from the well or creek and poured into the galvanized tub that was placed in the kitchen where hot water could be added. This was a lot more convenient than carrying the hot water to another room. It worked very well in the winter, the room was nice and warm when you dried off and dressed.

We always had a pail of fresh drinking water in the kitchen with a long handled dipper that we all used. You would think with all of us drinking from the same dipper that when one of us had a cold or flu, we would all come down with it, but Mother usually had our drinking water (or juice when we could afford it) poured and sitting by our bed. After we moved to Penrose, the folks managed to get beds for all of us.

The lumpy, used cotton mattresses were placed on springs on the floor or on an old bed frame. Mother and Dad slept on a feather bed. Feather beds were quite common back then, and they were every bit as comfortable as they used to say—you haven't really slept until you've slept on a good feather bed.

All of our cooking and heating was done on wood-burning stoves, but firewood wasn't all that easy to come by. One weekend when our wood supply was low, Bill borrowed a horse and wagon from one of his classmates and Dave, Lila and I got in the back of the wagon and rode up toward the mountains, following the ruts of what had once been a road. Parts of it were hardly traceable as we picked our way between rocks and brush toward what at one time been a forest of small pines. We stopped as we entered the scrub timber and gathered some of long dead—fallen branches and loaded them in the wagon and took them back down the long and winding trail home.

We had fun riding in the wagon it was a slow journey back down to the main road and home. Our bottoms were pretty sore from the wagon ride; but we brought back a pretty good-sized load of wood. We had no chain saw and the pieces that were too long to fit in the stoves were cradled between two sawhorses and sawed with a handsaw.

With seven kids it was hard for mother and dad to keep us all in decent clothes. Back then we made do with patched and hand-me-down clothes and shoes. But once we had signed up for relief, Mother found out that they also had a program where those parents that wanted to, could take classes in using a sewing machine, and practice with it by making clothes for the WPA workers. Mother took the course and learned to be quite a seamstress. She was able to make the girls' dresses and shirts for us boys as well as patch our worn-out clothes. She never lost her talent, she was still active patching and altering her clothes until her health started failing in her late eighties.

We always changed into our play clothes when we got home from school. One evening, Bill had put on his best clothes to go to a high school dance. His clothes looked pretty nice, considering, but his only pair of run-over shoes looked pretty rough; they were well worn and scuffed up. We had no shoe polish, but he decided he could darken them with soot that he got from the underside of the kitchen-stove lids. We always seemed to, as the saying goes, "make do."

We played outside after school most of the time, unless the weather was bad. If we had chores to do, they were done first, then homework, and then, we could listen to our favorite radio programs. We had no television, but I think we enjoyed listening to the radio just as much if not more; we could make better use of our imaginations. Some programs, like mysteries and scary stories, played on Sunday evenings. We also had comedies like *Blondie and Dagwood, Fibber McGee and Molly, Jack Benny, The Red Skelton Show* with his famous funny character Clem Kididdlehopper. We were allowed to listen to these after the folks had listened to the news broadcasts by either Walter Winchell or Lowell Thomas, two prominent news reporters of that time.

The Lone Ranger came on in the early afternoon when we came home from school. Some of the other programs we really enjoyed were *Inner Sanctum, I Love A Mystery* with Jack Doc and Reggie, *The Green Hornet, Mr. and Misses North, The Whistler,* and *Mr. District Attorney*. *Inner Sanctum* and *I Love A Mystery* were our favorites. These two always left us in suspense, eager to hear the next episode, eager to hear what happened to the characters–Jack, Doc and Reggie trapped in a cave or in some scary life-threatening situation. When those stories came on, the room became ultra quiet and anyone that coughed, sneezed, or continuously cleared their throat received some pretty hateful stares and words. We really hated it when thunder and lightning storms created static, cutting out parts of the story. It seemed to happen in the most exciting parts; the lightning would cause the station to fade in and out.

We enjoyed these programs even more when we went to visit our second cousins, the Whites. They had a nice big cabinet model, a Zenith or Philco; I'm pretty sure it was a Zenith. It had quality sound and more stations than our little radio. It had a small green light about the size of a marble that came on and we watched it glow brighter and brighter as it reached full power. When it reached full power it opened up like the pupil of an eye and was a deep emerald green. Listening to our favorite radio programs on those nice big radios was equivalent to seeing a big screen TV instead of a small table model. We not only enjoyed listening to our favorite programs, but we liked playing a Victor phonograph or a "Victrola," as it was called. It had a picture of a dog with his head cocked over to the side as if he were listening to the music as it came out of the curved trumpet-like speaker. The phonograph wasn't electric or battery powered; it got its power from us, winding up the turntable using the crank on the side of it. It was cranked after each record was played, depending on the number of songs on the record. It was also fun letting it completely run down while listening to the voices become more distorted as it ran slower and slower.

For other recreation, we played cowboys and Indians with homemade toys like rubber band guns and slingshots, and bows with arrows that never seemed to work. Dave and Bill were both good at carving. The rifles and handguns they made were some pretty real-looking stuff. They also made wooden sailboats that we sailed in the ditches and large puddles after a big rain. We played baseball and Kick the Can, kind of a poor kid's soccer, and. of course, Hide-and-Go-Seek.

Sometimes the neighbor kids would bring their bicycles over and the older kids, Dave and Lila, would take turns riding with them. Another thing we did to have fun, I should call it a sport all its own, was taking turns rolling down a hill in a car tire. One of us, if we were small enough, would brace our self inside the tire and let another kid roll it as fast as he could until he let it go rolling down the hill after a good start. The contest was to see who go longest distance before falling over. Doing this one time was enough for me after skinning my elbows and knees up pretty bad. Most of those old car tires of the late twenties and early thirties were narrow and very hard to brace yourself in.

When we had accidents such as stepping on rusty nails, getting a cut or getting skinned up, Mother doctored us up using turpentine and sugar. Turpentine was used to clean the wound and sugar to stop the bleeding. The wound was then wrapped with a clean cloth, which was taped to hold it in place. If we were out of tape, the dressing was tied on using strips of cloth torn from an old sheet, which was used both for the dressing and ties.

Turpentine, sugar and a clean, old sheet were the main things we had in our First Aid kit. This was the remedy for bleeding stubbed toes, pocketknife cuts, puncture wounds, and doctoring smashed fingers from hammer blows or from having a car door shut on them.

I started my first year of school in Penrose, Colorado. As in all schools at this time, the principal and teachers were well aware of our country's possibility of getting involved in the war going on in Europe. Those of us old enough to remember the times will never forget the news reports of the possibility of the United States going to war discussed among our parents. One of the songs we always sang in the little Penrose School was the song, "God Bless America." We sang this song after we said our Pledge of Allegiance to the flag.

The little Penrose School was the only place that I remember hearing the words leading into this song. These are the words as well as I can remember:

God Bless America

While the storm clouds gather, far across the sea

Let us swear allegiance to a land that's free

Let us all be grateful for the land we love

As we raise our voices to the stars above

God Bless America, land that we love

Stand beside her and guide her

Through the night like the stars from above

From the mountains

To the prairies

To the oceans white with foam

God Bless America, my home sweet home

God Bless America, my home sweet home

Like most kids, we enjoyed our pets, especially the different dogs we had. One in particular was Tippy, a little male rat terrier that Bill brought home one night after visiting a friend who was giving away puppies.

The first thing he did was chop off his tail; he said he looked too much like a sissy with his long tail. He laid it out carefully on the chopping block and chopped it off with a hatchet and Mother kept it doctored it up with turpentine and sugar for a couple of weeks until it healed. Tippy ended up with a tail about an inch long.

The little dog was just a little bigger than a Chihuahua. All of us kids loved playing with him and us boys were always teasing him. We teased him so much He would growl at the slightest hateful tone in your voice. He loved to be picked up and held; but you didn't dare to get pushy with him or you were liable to get bit. He slept at the foot of the bed under the covers and you didn't dare take your feet and push him out of his warm spot. He was one of the meanest little dogs we ever had. He was a good little watchdog and never made a mess in the house. We had other dogs throughout the years but little Tippy was our favorite.

When Aunt Minnie and Uncle Grant would bring the children for a visit they would nearly always bring their dog Big Paws, he and Tippy usually got along pretty well. Big Paws was a mix; he had short legs and a long body like a dachshund, and his head looked like a German shepherd. He got his name from his big paws, of course. Big would chase two-pound coffee can lids like dogs today chase Frisbees. He would lay the lid at your feet and wait patiently for you to throw it for him.

It seemed like one of us kids were always finding a baby bird that had fallen from a nest or a stray animal that had been abandoned by its mother. The baby birds with their broken wings or legs usually died after a few days of nursing and handling; but not so with a baby squirrel that Laura brought home. The squirrel with its bushy tail and big eyes was quite cute. She named it Wimpy after the comic book character.

It wasn't long until Wimpy was out of his box and scampering all over the house. When excited he would make a popping noise with his teeth, jump up on the back of the couch and into a string-less guitar Bill had hanging on the wall near the end of the couch. At this time, we had a yellow dog named Knobby. (His name came from his real short tail that looked like a knob.) He and Wimpy seemed to get along fine. Knobby, being a pretty old dog hardly did anything but eat and sleep. Outside, he would find a shady spot to lie where it was cool, and he would bark when any stranger approached the house; that was about all he was good for. When he was inside because of bad weather, he usually got as close to the potbellied wood heater in the living room as he could.

One cold afternoon, Laura had just sat down getting ready to warm her feet when she noticed Wimpy scampering about in the living room. At first she thought he was going to come up to her chair where she could pick him up. She sat real still as she watched him jump from the couch and start over toward the stove. Knobby was splayed out, soaking up the heat, when along came Wimpy. He cautiously walked to Knobby, checking him out, and all of a sudden with lightning speed, he clamped his teeth on poor Knobby's testicles. I guess he thought he was gathering nuts for the winter, as squirrels do. Knobby jumped straight up almost hitting the stove, yelping at top of his lungs, startling everyone in the house, before heading for another room.

One morning when Mother was getting ready to fix breakfast she got the surprise of her life. She had left the lid off after loading the stove the night before. With the damper open, she lit the paper under the kindling. When she saw that it was flaming up good, and had just started to put the lid on, out jumped Wimpy, causing her to drop the stove lid to the floor. Wimpy leapt to the floor and ran for the living room to his hideout in the string-less guitar. That afternoon, Laura was able to pick him up he was singed pretty close to the skin, especially his tail; but he was other-wise all right. Sometime later, she found out that Wimpy wasn't a male squirrel at all, but quite pregnant.

Bill always had something going; if he wasn't teasing one of us or the dog, he was experimenting with chemicals and firecrackers, making explosives from shotgun shell powder and blowing lard bucket lids high in the sky by mixing lump carbide and water, the same kind of carbide that was used in miners' lights, which they wore on their helmets. He would punch a small hole in the lid, put in a fuse made of twine soaked in kerosene, light the fuse, step back a few paces, and watch the lid as it was blown a couple hundred feet or more into the air. He just loved setting off firecrackers in anthills and bottles and cans.

Another one of his hobbies was catching spiders, insects, scorpions, snakes, and lizards. Tom, still in the crawling stage, was playing on a blanket in the living room when a rattlesnake that Bill had sitting on a shelf in his bedroom, penned up in a screened-in cage, somehow squeezed out the opening on the top. The snake dropped to the floor and slithered out of the bedroom. Mother had just come out of the kitchen when she saw the rattler in the corner of the living room. She grabbed up Tom and the blanket, quickly set him down in the front yard, grabbed a hoe that was leaning against the wall, and managed to get the snake out of the house.

When Bill came home, his snake was in pieces in the driveway. He really got chewed out. Needless to say, he never brought anything that dangerous in the house again.

First Bicycle Ride

One Saturday Bill brought home an old, skinny-tired bike that someone had thrown in the dump. The bicycle, even though it was pretty well beat up, could be ridden after a little work.

The two biggest problems were the tires and brakes; and, of course, the handlebars were bent and the bolt that was used to tighten the seat was stripped, and the seat would swivel all over the place, especially if you tried to pedal the bike standing up.

Lila was the one to think of it first. She threw away the rotten tubes and stuffed each skinny, cracked tire with rags and used bailing wire in several places to hold them on. It was not a bike you would ride to school or be proud to show your friends. Dave tested it out on a little hill not too far from the house, and he and Lila were taking turns on it when Lila talked me into trying it.

With the seat at its lowest, I was helped on and was able to reach the pedals. She pushed me around the yard, helping me learn to balance, and then decided to take me to the hill where she was going to let me try it on my own. My feet were placed on the pedals, or I should say, the center bolts, which were actually all there was left of the pedals. She let me coast down the hill holding on to help me keep my balance. She did this several times, letting me gain my confidence in being able to ride down the hill without her help. About the fifth time, I was ready to solo. What a thrill! She gave me a pretty good push to get me started. What a ride, I was doing it! I was on my own; she had let go of the swiveling seat and I was off!

I tore off down the hill on my white-knuckle ride, trying to dodge the rocks and potholes in what was once a road. About half way to the bottom of the hill, I started to waver and lose control. I had my foot on the pedal, pushing down as hard as I could trying to slow down, and all the while I could hear Lila laughing like her sides would split, as I was desperately trying to slow the bike down, keep my balance, and miss the rocks and pot holes. I swerved to miss a pothole and hit a good-sized rock, sending me cartwheeling down the hill. I guess watching me must have been hilarious. I went one way; the bike went another. I ended up on the side of the road rubbing my skinned elbows and knees. Lila was still laughing as I picked up the old bike, gathered up the seat, and started up the hill. The laughter caused my pain to turn to anger as I dropped the bike and seat in front of her. She had just started down the hill when I began pelting her with rocks. After she did a lot of apologizing and checked me out to make sure that I wasn't seriously hurt, we started all over again.

We also enjoyed our trips to the little country store. We hardly ever bought more than three or four items; our grocery list consisted of staples like coffee, flour, beans and lard and maybe a can of cream, and we nearly always took our two-gallon kerosene can to be re-filled. After one of the twins or Lila paid for the groceries and kerosene, we were allowed three or four pennies for candy or bubble gum, which we enjoyed on the way home. The candy was kept on the counter in big glass jars; it was a tough decision picking out the candy because of so many different flavors of penny suckers, hard candy and licorice sticks. They had peppermint, orange, grape, lemon, root beer, and butterscotch–my favorite. We especially loved the suckers with butterscotch and crushed peanuts. The bubblegum was fun too with its tiny comic-book waxed paper wrappings. The store was about a half mile from the house, and each one of us shared carrying the groceries and full kerosene can. Sometimes when we went for kerosene after a good rain, we would loosen the lid on the pouring spout letting a few drops land in the puddles on the side of the road so we could watch the beautiful rainbow colors that emerged. (My wife Jeanie told of a time when she and her brother Bud got a hard whipping for doing just that. She said they got carried away and wasted quite a bit of kerosene that was supposed to go in their lamps.)

One of the places rented in Penrose had a good-sized apple orchard and the landlord gave us a break on the rent if we would help maintain the orchard. One of the jobs was spraying the trees to keep out the worms. When spring arrived, Laura, Leta, and Bill sprayed the orchard. Bill drove, or I should say, towed the sprayer. The machine that he used to tow the sprayer was an old homemade tractor that had been assembled from a 1929 Ford truck with no cab. Towed behind it was an old homemade trailer with a large tank filled with a white solution of arsenic and water. The twins walked alongside the big tank and sprayed the trees with hand-held sprayers that had extended nozzles for reaching the higher branches. Bill would have to stop periodically to pump up the pressure in the tank. None of them wore masks; they just had cloths tied around their heads and cloths covering their noses and mouths. At the end of the day, they were covered from head to foot with the white spray. We thought they looked like ghosts. It's a wonder they didn't get sick from the arsenic mix they were spraying on the trees.

We had some nice neighbors when we were living in Penrose–the Derritos, the Thomases, the Molellos, and the Poncellos.

The Poncellos had dairy cattle and they kept us supplied with plenty of fresh milk; all we had to do was take our bucket up to the dairy barn at milking time. We took all they would give us. We had no icebox or refrigerator to keep the milk in, but none went to waste. When it started to sour, Mother put the sour milk in the churn and we made fresh buttermilk, skimming off the nice white butter that was delicious on our cornbread.

Mrs. Molello gave the twins each a pair of run-over, high-heeled slippers and some nylon stockings with runs in them. They were glad to get them; they were the only high heels and nylons they had.

The winters in Penrose, although not quite as severe as those in Cripple Creek and Box Canyon, were nevertheless quite cold. At night after the potbellied wood heater cooled down, the house got real cold. Many mornings we woke up to see ice in the water bucket as well as water frozen in the water glasses left on the table from supper the night before.

Mother, Bill and Dad were usually the first ones up and had fires going in both stoves, so it didn't take long to heat up the house. We didn't miss much school from having to stay home with sore throats and colds, but there were many days when we got up to get ready for school and the temperature was well below zero. We didn't have anti freeze back then and people that didn't have heated garages drained the radiators on their cars and trucks at night to keep the engine blocks from cracking when the temperatures dropped sometimes twenty degrees or more below zero. During the day, when the snow melted and ran off the roof, long icicles would be hanging from the eves, some almost touching the ground.

It was rough waiting to catch the school bus in the freezing weather. Our house was about a half mile from the bus stop. We had to walk down the lane to the main road to the bus stop. The country bus stops had no benches or shelters and sometimes, when the bus was running late, we would be running in place or stomping our feet and swinging our arms to keep our blood circulating, trying to stay warm until it came.

Bill suffered more with colds and tonsillitis than the rest of us. He would never wear a cap or anything on his head when he was out cutting wood, walking or playing in the snow. Mother was always getting onto him about it and doctoring him for a sore throat. She usually used Vicks Vaporub and cough drops when we could afford them. One of her home remedies for sore throat was lard mixed with kerosene. She smeared the lard and kerosene on your throat and wrapped it in a flannel cloth or soft sock and after heating it up near the stove, she pinned it with a safety pin. The kerosene-lard mix loosened the congestion, taking the soreness away. We always hated to be doctored up that way, having the smell of kerosene on our clothes and bedclothes. We had no washing machine; our clothes were washed in a galvanized tub using scrub board and lye soap.

Other home remedies we took to keep us healthy were doses of castor oil as a laxative and a spoonful of sugar and turpentine to keep us from having worms.

On those cold winter days and nights, we popped our corn in an old cast iron skillet and made "snow cream" from the fresh falling snow. Laura and Leta, and Lila, the youngest of the girls, would scoop the freshly fallen snow from the eves of the roof and fill our big bowls with the fresh, clean snowflakes, and add sugar and vanilla, making a poor man's ice cream. We also crunched on the long icicles from the eves of the house. No one in his right mind would think of making snow cream now; they would really be taking a chance with all the pollution in the atmosphere.

When we couldn't be outside, we drew funny pictures, played cards, and listened to the radio if the static wasn't too bad. Lila was very talented when it came to drawing comic book characters. She just loved drawing Bugs Bunny and Elmer Fudd scenes.

Bill was good at playing the guitar and we all would gather in the warm living room and sing; the twins, Mother, and Lila were really good at harmonizing. We would crack up laughing at the twins and Lila when they would clown around, dressing like someone they had seen in a play or in a Charlie Chaplin movie. We all thought Leta was the funniest. She must have been double-jointed. She had a way of throwing her knees out of joint; instead sticking out, they would drop back the other way giving her legs from her knees down a backward bow.

Our lives were pretty simple in those bittersweet days. We never had much in the way of material things, but we had each other and we always did things together. I guess that is what mattered the most.

Bill kept us laughing in the worst of times. No matter how tough things got, he never lost his sense of humor.

All the times when Dad was struggling to find steady work, Bill was always the one we looked up to. When Dave, Tom or I had trouble with bullies in school, we told our big brother Bill, and he took care of the problem. With Dad away from home so much, he was more of a father figure than Dad.

He was quite a storyteller, too. He talked very slowly with a serious expression on his face, making up the stories as he spoke. He captivated his audience; they would be spellbound until the end, when they realized they had been taken in by a big but funny lie.

Of all the kids in the family, Bill was the most mischievous. He was always pulling a practical joke on someone in our family, or on one of our cousins, or on one of his friends. I remember one story in particular.

Bill's Toilet Bowl Ride

It was a warm and sunny day, a nice day for moving from the heavy stuff from the Cripple Creek High School gymnasium to a storage room in one of the old brick buildings downtown. Bill had made several trips with a flatbed truck, hauling various things from the school to storage. The gym had been remodeled along with several of the schoolrooms. Three boys rode on the back, Bill and two of his classmates, to make sure that nothing bounced off into the street, and to help unload and sort out the stuff when they got to the storeroom.

Among the various boards and pipes and sinks, they had placed an old worn-out commode. Just before the truck pulled out of the schoolyard, Bill lowered his pants and sat on the commode, assuming the position of using the toilet, as if he were doing his "business."

The two classmates who were sent to help unload were laughing so hard they almost fell off the truck as they slowly drove down through town with people gawking, pointing, and laughing in amazement as they watched them drive by. Anything for a laugh—that was our older brother Bill.

He was a bartender on and off from 1946 to 1980– the perfect job for a good storyteller. He laughed and joked all his life, clear up until his death in 1980.

♪

Our Christmases became much more memorable after Dad started getting part time work. Tom, Dave, and I talked about them years later. Mother and the twins always fixed such wonderful delicious dinners, includ-ing homemade cakes; and they managed to fill our big wool stockings that hung on the wall near the Christmas Tree with homemade candy and store-bought candies like striped peppermint, and other, different colored Christmas candies, along with an orange, an apple and a handful of unshelled walnuts, peanuts, and pecans.

Our Christmas trees had no strings of lights and had very few, if any, store-bought decorations. Instead, they were decorated with popcorn balls and red and green homemade paper chains. We enjoyed homemade walnut fudge, the "divinity" or white candy, and peanut brittle. Peanut brittle was one of Mother's favorite candies, which she and the girls made the night before.

Our Christmas presents were usually clothing, socks, and underwear, and a few toys, maybe one each for Dave, Tom and I. We didn't always get the toys we had dreamed about, toys that we had seen in the catalog we read in the outhouse; but we were happy just the same. Of course, we sometimes got into fights if we got three toy guns that were similar, and one was just a little better than the others. We got so we marked our toys to keep track of who owned which toy. This kept down the fights.

Mother was real strict with us when it came to morals and honesty. She made it a point to check out who we played with and we got a good switching if we were caught telling her a lie. She could tell right away if we had invited someone in that would be a bad influence on us, and as soon as they had gone, we were told whether or not we could have him or her back again.

One day I was playing in the backyard when I looked over into our neighbor's yard and saw some shiny, big nails—a few on the ground and a handful on their porch. There was no one around so I took a handful of them from the porch and started pushing them into the ground, making a little fence out of them, fencing in the little cars I was playing with.

I was busy building my fence when Mother came out the back door to check on me. She asked me where I got the nails and I showed her where I had taken them from the pile on the porch. She asked me if Mrs. Thomas had given them to me, and I told her that I had just taken them from the porch.

I knew right then I was in trouble. She had me go inside into a bedroom away from the rest of the kids and gave me a hard whipping, and made me take the nails back to Mrs. Thomas.

With my behind and legs stinging from the switching, and tears running down my cheeks, I knocked on Mrs. Thomas's door. With tears in my eyes and Mother standing behind me, I handed the nails to Mrs. Thomas, telling her how sorry I was for stealing her nails.

One day Mrs. Thomas brought over a nice blue suit. She said she was going through a trunk she had in the attic and had come upon this nice suit that one of her sons had outgrown She thought it might fit me. Mother thanked her, and after she left, I tried it on and it was a perfect fit. I was really proud of my suit.

About a week later, Mother and I were walking down the road and I was wearing my nice blue suit when Mrs. Molello, a neighbor working in her yard, spoke to us as we walked by.

"Oh Jimmy, you look so nice, " she said. "Where did you get that pretty blue suit?"

Of course I didn't lie.

"Mrs. Thomas gave it to me," I said.

Later Mother was telling Aunt Minnie about it and said it embarrassed her when I had told Mrs. Molello that Mrs. Thomas had given it to me. Mother, the poor but proud person that she was, didn't like all the neighbors to know that people were giving us clothes. She and Dad were both that way–proud and inde-pendent. But after all, I was being truthful about it.

Trip to Canon City

When I was a just a little over five years old, Bill took me on an unforgettable trip to Canon City and Florence, Colorado. Both Canon City and Florence are on the banks of the Arkansas River, which flows down out of the mountains into the famous Royal Gorge. Canon City was mostly supported by people who worked at the Colorado State Penitentiary.

On that trip, I got to see a black person up close for the first time while I was waiting for Bill to come back from inside of a store. Back then grownups didn't worry about leaving a kid or dog in the car while they shopped unless it was in heavy crime area. I was sitting in our Chevy when a black man walking down the sidewalk almost touched the side of the car. I had seen black people before from a distance, but I had never seen one up that close until that day. I had been looking at the bright traffic lights on the corner across the street when he walked by. He really scared me.

When Bill came out, I told him right away what had happened, and I told him that he scared me when he came so close to the car. Bill explained to me not to be afraid of colored people, as they were called at the time, and that they were just like white people—there were good ones and bad ones—and there was no need to be scared.

We left that store, and then stopped at a grocery store, and Bill went in and picked up a sack of tobacco, some cigarette papers and a candy bar for me. After he rolled and lit a cigarette, we drove around town and then headed back toward Penrose. I had all kinds of questions about black people to ask Bill on the way back.

As we were headed back, we came across one of the longest freight trains I have ever seen. The train was coming from Colorado Springs or Denver, making a run to Florence and Canon City. We could see it coming from far off in the distance as the billowing puffs of snow-white steam puffed up from the big smokestack. Bill pulled the old Chevy over on the side of the road and rolled another cigarette as we watched it, some of it disappearing as it weaved its way between the rolling hills. I guess it was the cold clear Rocky Mountain air that made it such a pretty sight as the clear freezing sky turned the puffs of steam into small silver clouds.

That same winter of 1939, two of Dad's friends from Kentucky stopped by to see him. What a surprise it was when they pulled up in front of the house in an open cab Model T truck with no top, and just a mani-fold heater to keep their feet and legs warm. We couldn't see how they managed to keep from getting frostbite as cold as the weather was.

We had no idea who they were or what they wanted until Dad walked to the door to greet them and called out their names. They stepped inside and he introduced them to us. I do remember their names—Howard Reddick was the driver of the old truck and the other man was Arch Manning.

Dad had written them a couple of months earlier and told them that they had a good chance of going to work in Pueblo at the smelter, where the ore was melted to create copper.

When they stopped by the house they didn't stay long. Just long enough to have breakfast, which was plenty of coffee and some of Mother's mouth-watering biscuits and gravy. They stood in front of the potbellied stove and got good and warm while Dad gave them directions for how to get to the smelter near Pueblo.

We kids watched as they walked out through the snow and cranked up the old Model T truck and chugged off down the road sitting up stiffly in the open cab. The sun was shining but it was freezing cold. We stepped outside and waved till they were out of sight. Pueblo was over a hundred miles south of where we were living. We knew it would be a tough, freezing ride for them even though they wore big, heavy overcoats. I'm sure they stopped at different places to warm up along the way, but traveling that far in that old truck with an open cab with only a manifold heater must have been one freezing journey. In those days, the days of the Model T, the cabs of trucks and the cars of those years were heated by the heat coming from the exhaust manifolds while the engines were running.

Denver

We left Penrose in the early summer of 1941. I had finished my first year of school and I was seven. Dad had finally got a decent paying job at Remington Arms in Denver.

The move to Denver was great. Dad had steady work and we moved to 533 South Stewart Street, a house within the city limits, a house with electric lights and indoor plumbing. No more walking through the snow to the outside toilet and no more carrying water from the creek or drawing it from a well.

Dave and I were so fascinated with the flush toilet; we must have flushed it a hundred times in the first two weeks after we moved in. Mother was always onto us, or telling us to quit flushing the toilet so many times and wasting so much water. We even had a bathtub! It was one the nicest houses the family had lived in since leaving Detroit. Dad traded the old, dark blue '28 Chevy for a brown 1934 Chevy four-door. We were financially better off than I could ever remember. We even had money for nice school clothes and shoes.

I especially loved my nice, new (not hand-me-down) coat and lined boots that I wore to Garden Home Elementary school that winter.

One night after we finished our shopping in downtown Denver, Dad decided to drive around the city before heading home. I'll never forget seeing the bright lights that lit up some of the billboards while we kids enjoyed eating bananas and candy as we cruised up and down the street.

One big billboard sign that stood out, the one we all most remember, was a Negro nanny, as she was called in those days, who was dressed in red with white polka dot clothing, including a polka dot bandana on her head. The sign was a scene of her scrubbing clothes on a big washboard that was sitting in a bright galvanized washtub. It was a sight to see as we watched her moving her arms up and down as she scrubbed the clothes.

Garden Home grade school in Denver was bigger than any school any of us had ever attended. We bought our lunch at the school cafeteria. We had good food, milk, and dessert; and we had it all. We enjoyed the good food, but really didn't like the big school all that much. It was quite a switch from the little country school in Penrose and the fairly small schools that the bigger kids had attended in Williamsburg and Cripple Creek.

Dave and I liked the school in Penrose better. In Garden Home, there were just too many rooms, too many kids, and too many things to try to remember. We had trouble finding our classrooms when we came from the assembly room or cafeteria.

We really looked forward to going home after school to listen to our favorite radio programs without a lot of static—the static that interrupted them when we lived in the mountains and in Penrose.

Dad worked nights and Bill drove us to school in the family car. When the snow started to fall, we sometimes had to shovel the snow out of the driveway and help Bill by pushing the car as he pulled it out onto the street. After living in Cripple Creek and Penrose, the heavy snows and ice didn't bother us.

We hardly knew any of our neighbors. We didn't really associate with any of them, just an occasional hello if we saw them getting in their car or saw them in their yard. When the weather was good, we spent most of the daylight hours playing outside.

One afternoon Dave, Lila and I were playing ball in the front yard. Lila threw the ball to Dave and he missed catching it and the ball rolled across the street into the neighbor's yard. When Dave picked it up and started back, the neighbor came out of the house and yelled at him to stay out of his yard. I guess after hear-ing us running in his yard several times, he was going to put a stop to it. He ran over and grabbed Dave and started shaking him, swearing and threatening him. He didn't realize he had made a big mistake until he was struck just above his glasses with a sharp rock.

Lila had been watching from across the street and no one, especially a stranger, was going to hurt her little brother. She picked up a rock and let fly with her deadly aim; she laid the man's head open. She then ran across the street and started pounding on him.

With blood running down his face while being punched and kicked in the shins, the neighbor he loosened his grip, and Dave and Lila ran home. All the noise and commotion brought Dad out of the house; he walked over and told the neighbor that he would be in big trouble if he ever laid a hand on one of us again.

We were living in Denver when the war broke out. I'll never forget that day when all the grown-ups were bunched up around their radios listening to the news about Hitler's armies invading one country after another in Europe, and now the Japanese had bombed Pearl Harbor. Lowell Thomas, Walter Winchell and others were broadcasting the reports. Lowell Thomas nearly always started out by saying, "Good evening to you, ladies and gentlemen, and all the ships at sea. This is Lowell Thomas bringing you the latest news." The famous news reporter had spent his childhood in Cripple Creek.

After the attack on Pearl Harbor, people all over America were enlisting in the armed services. People in all parts of the country were trying to do their part in the war against Japan and Germany. Men and women, old and young alike were enlisting or trying to enlist in the armed forces. Teenagers were lying about their ages or health problems to get in. Those who didn't make it went to work in defense plants, aircraft factories, and shipyards.

Bill was seventeen when he enlisted in the Navy, along with several of his school classmates. All of us were in tears when he left for boot camp in San Diego in the summer of 1941. None of us had the slightest idea where he might be going after boot camp. We had no idea that he would be serving his country as a gunner on old Merchant Victory ship.

Bisbee

By this time, the Carrolls had moved to Bisbee, Arizona and it wasn't long after that we moved there too. It was another old mining town like Cripple Creek, about ninety miles south of Tucson and about eighteen miles from the historic town of Tombstone. Like Cripple Creek, it was built around the hillsides surrounding the mine. Mining for copper was what they called "open pit" mining. The areas around Bisbee and nearby Lowell were big open canyons where the copper ore had been taken out with huge earth-moving equipment and was then loaded on rail-road hopper cars and transported to the smelter in Douglas.

Coming into Bisbee from the east is the little town of Lowell. Like Bisbee, it also overlooked the open pit copper mine. When we first got there, we rented a place in Lowell. Dave, Lila and I had just started school when Dad got a job at the smelter outside of Douglas, a town right on the Mexican border. Shortly thereafter, we moved again.

We moved to a farming community in Sulfur Springs Valley called Double Adobe, which was not far from Douglas; Mother was able to raise a garden and both of the twins got jobs working at Douglas Army Air Base. We all really enjoyed the little farm. Our closest neighbors, the Shaw family, had several children, and it didn't take long for us to make friends. Both farms got their water for drinking and irrigation from wells pumped by windmills. And both places had ponds. Our pond was mostly filled with cattails and plants and plenty of bullfrogs. The Shaw's pond was much bigger and was used as a water hole for their cattle. It really wasn't a place to wade or play in. The pond was about three feet deep, one foot of water and two feet of thick black mud. It was loaded with catfish, the ones with the needle sharp dorsal fins.

One day Dave and I were playing around the pond when we found a big pan that was used for mixing straw with adobe for brick making. It was about six foot long and four foot wide and was made like a big, open tin box with curved ends. It was big enough for two men to stand in as they tromped the straw, mixing it in with the adobe and water. We decided to use it as a small barge and pole it across the big muddy water hole. We went back to the house and picked up a couple of skinny, mesquite poles that had once been part of a fence and went back to launch our barge. We put in our poles and managed to drag it over and slide it into pond. After checking it for leaks we stepped in and started poling it across.

We were about halfway across when our barge sprung a leak and quickly settled on the bottom in the mud, and we had step out and walk to the shore. We both yelled in pain when our bare feet out came down on the ice pick-like prongs that stuck out of the dorsal fins of the dozens of catfish covering the bottom of the pond.

By the time Dave and I reached the shore, we both had several black puncture wounds in the bottoms of our feet from the fish and black mud, and did they hurt! Stepping on those fins was every bit as bad as stepping on a nail.

None of us will ever forget the Shaw family, es-pecially Stoffer Shaw, the father. He was definitely the ruler. He never invited us to eat with them if we happened to show up at mealtime. Stoffer would tell his kids to leave the table one at a time when he decided they had eaten enough. He would tell them to finish what was on their plate and get up from the table. We felt sorry for the kids; we had never seen anything like that before. It wasn't that they didn't have enough food either. There were four children, three girls and one boy. The three girls were Virginia, Erlene, and Tootsie the youngest. Doyle was their only boy.

Tootsie was a real skinny, little freckle-faced girl that Bill and my sisters were always teasing me about.

We had a windmill, a tank house and a small pond that was fed from the overflow from the big wooden tank that sat on a platform about twelve feet off the ground. The pond was covered with lily pads, and bamboo and cattails lined the banks—an ideal place for frogs. Underneath the tank house was a small room with a shower. It was our recreation room and our playroom. It was an ideal place to get out of the blistering Arizona sun. The little room had a small card table, four beat-up chairs and a small bed. There we read comic books, played cards, and whittled our toy guns. We had only one BB gun. It was the popular lever action Red Ryder BB gun, which Dave, Tom and I shared with the Carroll boys. When we ran out of BBs, we dropped wooden matches down the barrel for ammunition.

One afternoon when Dave was taking a shower, I pulled back the curtain and let Dave have it in the butt with a match. There was no one in the tank house but Tom, Dave and I, and by the time Dave jumped out of the shower and got his clothes on I was long gone, leaving the BB gun on the bed. I didn't get a whipping, but I did lose the privilege of using the gun for a long time.

The Perilous Flight of Captain Midnight

A big part of our fun time was spent playing cowboys and Indians; we tried to mock the gunslingers we had seen in movies. One of our favorite movies, a series that played every Saturday afternoon, was Captain Midnight. The movie was a real thriller and always ended with Captain Midnight facing some impossible, breathtaking danger like being trapped in a cave that was filling with water, or in some other dangerous predicament like being strapped to a table about to be sawn in half by a monstrous blade.

I was always trying to play Captain Midnight's heroic part. For my birthday, the twins gave me a special present they had bought at a surplus store; it was just what I needed to complete my Captain Midnight costume. They brought home a wool-lined aviator's helmet with the big goggles. Wearing the helmet, goggles and an old cut-off brown bathrobe, I was outfitted with the perfect Captain Midnight costume. In the blistering heat, you could see me battling the Forces of Evil as I dashed between sheds with my helmet on, goggles down, and my cloak flying in the wind as I ran, sometimes jumping off the roof of a shed or from the iron wheels of an old, broken-down, iron-wheeled Fordson tractor.

There was an old homemade cement mixer in the back yard. It was an octagon-shaped wooden barrel supported by two large posts with a three-inch pipe that ran through the center. The center pipe ran through bearings that were mounted in the posts. On one end were a geared hand crank and a pulley for running the mixer with a gasoline-powered motor. The mixer could be powered by hand.

One day I was talked into climbing into the mixer for a little ride (a ride like being in an airplane out of con-trol). I was perfect for the ride. The door was too small for the other kids to enter, and after all, I was Captain Midnight, wearing my cloak, aviator's helmet, and goggles. I was helped into the cockpit and lowered down, straddling the three-inch pipe as the hatch was closed. I wrapped myself around the pipe as the barrel began to roll. It was cranked faster and faster, the kids making background music for my perilous flight.

After a few minutes, my crotch started to ache, my eyes were stinging from sweat, and my mouth and clothes were filling with debris from the mixer. I was ready to get out. Of course, the more I yelled to stop, the louder the background music became. When my cries became shouts and the mixer was then cranked to the stop position, I could hear the kids laughing. The more they laughed, the more enraged I became. As the pin was put in to keep the mixer from turning, I was clumsily trying to unwrap myself from the center pipe, as the door was flung open. There was no one to be found as I lowered myself to the ground, rubbing my crotch with burning sweaty eyes, my cloak and aviator helmet covered with debris from the mix-er. Anyone close by, especially the one doing the cranking, was in for getting hit with a rock or any-thing I could get my hands on.

I saved that nice wool aviator's helmet and big goggles and wore them when I rode my first motor scooter in 1948 when we moved to Burns, Oregon.

We also had fun with little planes, balsa gliders, which came in boxes of Pep, a wheat cereal that was popular back then. I remember saving several Pep box tops and sending off for their special offer of cockpit con-trols. I waited anxiously for three weeks for the order to come in. I was constantly checking the mailbox, expecting a box with a control panel. I was really let down when it finally arrived; it was nothing but gauges drawn on a piece of cardboard with red and black paint.

Cousin Grant made warplanes from balsa wood and tissue paper, and we enjoyed watching when he flew them. The planes he built were much more intricate and required a lot more work and patience to get them together. He painted with airplane dope and detailed them out with decals and all the proper markings, and they were powered by rubber bands.

♪

Sulfur Springs Valley was ideal for raising vegetables if you had a good well with plenty of water. Mother raised some of the best gardens. We always had plenty of fresh vegetables, and she did a lot of canning and dried beans by placing them on the low tin roof of the back porch.

We never had to worry about a lot of traffic on the country road that ran in front of our place. Just about the only people who used it were farmers and ranchers, the postman, and the school bus driver.

About once a month the Watkins Man would pay us a visit. He had all kinds of things in his van from brooms and brushes and liniments to soft drinks and candy. He would pull up near the driveway, get out of the van, and start in with his sales pitch as the women and kids bunched up around the van. Mother nearly always bought something. Most of the time it was a brush or a broom, and she bought us what he called "Summer Drinks for the Kiddies." They were bottles of lemon, orange, or strawberry flavored concentrate to be mixed with water. She also bought a variety of Kool-Aid.

Not long after that we said our tearful goodbyes to Grant when he headed for boot camp like so many others his age. In spite of how busy we were after he left, we all worried about Bill and Grant. Bill had become a Gunner's Mate in the Armed Guard, a name given the sailors who made up the gun crews on the old, slow-moving merchant ships. It seemed like every time we heard the news it was bad.

Aircraft carriers were being sunk, merchant and troop ships were going down with all hands after being torpedoed by enemy submarines, while hundreds of army soldiers and marines were being slaughtered in Europe and on the islands of Guadalcanal, Saipan, Okinawa, and the Philippines.

In the newsreels we saw when we went to the movies, we saw film after film of burning ships and sailors being pulled from a sea of burning oil and flotsam. Some were pulled out of the water with nothing below their life jackets; they had been half-eaten by sharks.

We worried about Bill more each day, as Mother searched the paper to see if Bill's ship, the M.V. (Merchant Victory) Day Star had been sunk. This ordeal–not knowing and not receiving any mail from him–really caused Mother to worry. It wasn't very long before she became sick. She got so sick; she couldn't keep anything in her stomach and started throwing up blood. Dad and the twins took her to Douglas to the hospital, and found out that she had an ulcer that had perforated her stomach causing it to bleed.

Aunt Minnie brought her children and came out and stayed with us the two weeks that Mother was in the hospital. While in the hospital, she had several transfusions. One donor was a soldier, a full-blood Cherokee Indian from the air base. She liked that, being one-quarter Cherokee herself.

Aunt Minnie was a real help. She was so much like Mother. She did all the chores, the cooking, the washing, and even the canning. The cousins kept us occupied and kept our minds off of worrying about Mother. Aunt Minnie disciplined us just like we were used to, except she didn't whip us quite as hard.

One day she sent Lila out to the tank house where Mother had fruit jars stored. Aunt Minnie was right in the middle of canning tomatoes when she ran out of jars. I, of course, was where I shouldn't have been; the windmill had stopped and I was standing on the windmill platform with a pocket full of rocks, throwing rocks at different things in the yard. When Lila came out to get the jars, she saw me and told me to come down. When she started across the yard, she saw that I was still up there, and she yelled up that she was going to tell Aunt Minnie. I flew mad and threw a rock at her, never dreaming it would hit her in the head.

The rock struck the back of her head, just above her ear, causing her to crumple to the ground. When I saw her fall, I began climbing down as fast as I could to see if I could help her. Just as I started toward her, Aunt Minnie, after hearing the commotion, ran out and found Lila sitting, crying, and holding her bleeding head. I knew I was in big trouble and ran and hid behind the cane and cattails at the edge of the pond.

After tending to Lila, she came looking for me, standing at the edge of the pond, calling for me to come and see what I had done to my sister. I caught up with her as she walked back toward the house.

I didn't get the hard whipping I so much deserved, but I did get a good lecture. I guess she felt sorry for me as I apologized to Lila, giving her a big hug, telling her how sorry I was for hitting her in the head with a rock.

I thought I had been invisible hiding in the tules wearing a light green shirt, one that Mother had made; but I'm sure that she saw me there in the bamboo and knew I was too scared to come out.

Before the summer was out, we moved to a nicer house. The family really liked the place. It was a three-bedroom, built in the thirties. It wasn't as nice as the one in Denver; but a lot more modern than the one we were moving from, and it came with five acres. The house had been well kept, but needed some repair, the worst of which was a crack in the foundation on one corner of the front porch, causing it to sag. Other than that, and after a few minor repairs, we were pretty well pleased with it. Dad had a pretty secure job and was making a decent wage. Dad, Mother and we kids managed to build a small barn, a pigpen, and a chicken coup. We had several laying chickens, a rooster, a couple hogs, and a Jersey milk cow. The hogs, a boar and sow, we were able to buy from the Harshas–the people we bought the house from. They had a hog ranch and raised acres and acres of produce.

It took a lot of feed for several hundred hogs and Bill Harsha, the owner would help out with his feed bill by hauling slop and mixed stale breads he got from the garbage cans in the alleys behind the restaurants and grocery stores in Douglas like uncle Grant had done in Cripple Creek.

Bill Harsha, a wino, was in his late fifties or early sixties and had a mouth full of long black decaying front teeth. He always wore long sleeved shirts or an old black leather jacket covered with grease and grime from handling garbage and barrels of slop.

Every day you could see him on his way to Douglas in his 1938 Dodge pickup with the high sideboards and fifty-five gallon drums he used for the wet stuff, slop. With the slop and the corn he fed, them made some pretty fat hogs and kept him with a full time job.

Sometimes when he came back from a garbage run he could hardly stand up after sipping wine all the way back from Douglas. He had two sets of gates, two pens. The first pen was to keep the hogs from going out on the road or into our yard if any got by him when he opened the gate to the main big hog pen.

He would sometimes be so drunk on wine he would stumble and fall when he reached up to open the second gate. We could hear the hogs squeal as he fed them; they would bunch up and fight when he dumped the barrels into the troughs. There were many times he got in the pen with them and at times climbed on the backs of some of the big boars, riding one like you would a pony.

Getting in the pen with that many hogs was very dangerous; we expected anytime to hear screaming as he was being eaten alive.

2 ON THE MOVE AGAIN

After mother got out of the hospital, we moved into a bigger house in the little town of Elfrida that we were able to buy from Bill Harsha, a hog rancher and produce farmer. It was much nicer than the one we had been living in when we first came to Sulfur Springs Valley.

We didn't have a windmill or tank house on the property, but we had plenty of nice clear water in the irrigation ditches to cool us off when we played out in the blistering heat.

McNeal was another small community about six miles south of Elfrida, which we only visited for gas or a pack of cigarettes for Dad or a can of snuff for Mother. It wasn't big enough to do any real shopping. Elfrida did have an attraction for us kids; not far from the center of town near the highway where the road forked, there was a tower that held a big beacon or searchlight. The right fork went to Bisbee, and the left to the highway, the route to McNeal six miles south, and on south to

Douglas. With the airbase not far away, it was turned on every night and when it made its search, it could be seen for miles. It would search the sky and then shine down low, just above the farms and ranches.

When we were living in our first place with the windmill and tank house, we would sit on the tank house platform and watch it make its sweep, and when it lit up the garden, we imagined seeing all kinds of weird things like the shadow of someone running between the rows of corn or people standing under the mesquite trees. After listening to so many scary stories on the radio, our imaginations went wild.

Lila used to get a big kick out of scaring Tom and me. Sometimes at night, when we would be getting some-thing out of a bedroom closet, she would sneak up behind us and shut the door, holding it shut while telling us in a scary voice that there was a ghost in the closet with long bony fingers that were reaching out to get us. When we were little guys, she scared us pretty bad. Even as we grew older we were always expecting to see ghosts. When Mother caught Lila doing this, or when she heard the terrified screams and crying coming from the closet, Lila usually got a whipping.

Our for-real, biggest worry was of being bitten by a rattlesnake or Gila monster, or by a black widow spider; or to be stung by a scorpion or find a centipede in one of our shoes. We were always watching out for our dog Tippy; he was with one of us most of the time when we played outside. We never let him wander off into the mesquite where he might get snake-bit.

♪

The Carrolls' house in Bisbee was built on the side of a hill, and like most of the neighboring houses, it had no real front or backyard, just the steep side of the hill. To get to their front porch, you had to climb several flights of stairs leading up to the little house, and the house itself was built on top of boulders. The balcony-like back porch, which had no room at all, was used for hanging laundry out to dry and as a place to hang mops and brooms.

When they came out to visit us, it was a real pleasure for them. They had our big yards to play in and plenty of space for playing ball, or riding bikes up the gravel road that had very little traffic.

They sometimes stayed a couple of weeks; that was about all Mother could take, having to watch after all of us, break up quarrels and fights, and doctor our skinned knees and elbows. They minded her pretty well; they were disciplined the same as us. She had no trouble giving one of them a whipping if he or she deserved it. Aunt Minnie did the same to us when we stayed at their house.

One day Dave and Grant were looking through some stuff in one of the sheds when Dave found a small wooden box on one of the top shelves. There was nothing else on the shelf but an old worn out hoe with a broken handle. When he handed the box down to Grant, he had no idea what was in it. They opened the lid and saw the box was almost full of what looked like over-sized, empty, copper .22 shells. As soon as Grant saw them he knew what they were. They were dynamite caps. He had seen some like them before at a friend's house whose father worked in the mines.

Grant decided right then that he must get rid of them to keep anyone from getting hurt. He decided the best way to do that was to build a small fire, throw in the caps, and throw some kind of cover over them before they exploded. Most people living in the country disposed of their own garbage and junk. They would usually find a place away from the house for their dump, which was a sinkhole or dry creek bed or wash. There was one behind some mesquite trees not far from the house. Grant and Dave searched the dump and Dave came back carrying an old, rusty washtub.

All of us kids followed them quite a ways from the house where they built the fire, making it small so the tub could cover it. As soon as the fire got going good, Dave tossed in the caps, and then ran as fast as he could toward the shelter of the mesquite. Grant dropped the tub over the fire and was soon right be-hind him. We didn't have long to wait for the caps to blow. It sounded like a gunfight when they went off, hitting the rusty, old tub. After everything had cooled down, we checked out the tub, noticing several holes where the caps had blown out through the rusted spots.

When the grown-ups in the house heard the noise, they thought Grant had set off firecrackers that he had brought out from Bisbee. Once again, we were very lucky that no one was hurt.

Bus Rides to Double Adobe

The school Dave and I attended in Double Adobe in 1947 was several miles from Elfrida and we had to get up early to catch the school bus. We gathered our books and lunches and walked down the lane to the main road to get to the bus stop. Sometimes one of us had to run back to the house to take the dog back when he followed us, or for something we had forgotten, like a schoolbook or the candy money that we got every Friday.

When we boarded the bus, about a half dozen kids already had their seats, but we still had plenty of seats to choose from. We usually sat in the middle on the right side; that way we could see the kids boarding bus and see if they were carrying comic books. We weren't allowed to read comic books in school, but we would take them to trade on the bus on the way to school or on the way home.

In the hot weather the bus got pretty warm and the windows were opened for ventilation. The school had no cafeteria; everyone brought their lunch. Most of the lunches were in paper bags and as the bus filled with kids, the air inside the bus was filled with the smell of peanut butter and grape jelly sandwiches.

There was one black kid in the first grade that brought his lunch in a two-pound bucket like the tapered bucket that lard came in, only his lunch bucket had colorful pictures of George Washington and old glory, and bright stars framed the picture of Washington on the lid. We had never seen anything like it before, and I haven't since. He would doze off as soon as the bus started to roll and be sound asleep when the bus pulled into the schoolyard. He only attended school at Double Adobe for a few days. We were told after he left that he had been transferred to Fort Huachuca in Arizona.

Fridays were our candy days. At recess, one of the teachers would stand on the back step and ring the hand-held candy bell and everyone would run from the playground and line up to buy candy. The candy was mostly chocolate bars, Milky Way bars and packages of Walnettos. Walnettos were individually wrapped little pieces of candy about one-inch square and a quarter-of-an-inch thick, made with caramel and crushed walnuts. All the kids really liked them. They lasted longer than chocolate bars and could be enjoyed on the bus ride home.

Mr. Cunningham, our bus driver had lived in the Double Adobe for quite some time. Driving by his ranch was spooky to some of us. The ranch had a family graveyard on the property not far from the house. The wrought iron rail around it could be seen from the road.

We were told that the caskets were in a room above ground, a mausoleum like those in some Catholic cemeteries. Of course, after telling Lila about it, she had us believing that the place was haunted and full of ghosts.

Mr. Cunningham was a nice, polite bus driver who didn't mind us eating candy or chewing gum on the bus. He did get mad, and sure let you know it, if he saw you throwing your candy and gum wrappers on the floor, or if he caught you tacking your gum to the bottom of a seat, or getting out of your seat while the bus was in motion. We weren't too fond of him, but he did run a pretty orderly bus.

The Wild Ride of Der Fuehrer

On the property next to ours, there were several burros. We could pet them and feed them; but, there was only one we could ride, and he wasn't too keen on that. It wasn't that he would try to buck you off; he was just a slow mover, even on his best days. We rode him around the yard and up and down the road. He would trot but never reach a full gallop. Mother never worried about us when we rode it. Maybe it was the hot weather; whatever it was, we could never get him to move very fast.

One day we were all taking turns riding him when Dave and Grant rigged up an old two-wheeled cart that they had found in of one of the sheds. They had trouble getting the old, dried-out, weather-cracked harness to fit; but, other than that, everything went pretty well. We took turns riding the iron-wheeled cart, pounding up and down the lane while our cousin Nancy was practicing her "Sieg Heils" and Hitler sa-lutes.

She had taken mascara and painted on a little, square, black moustache and had drawn dark circles under her eyes. After slicking her dark hair down just above her right eye, she made a pretty good impression of Adolph Hitler.

In the hot summer weather the burro began to slow down even more after the first few trips taking us up and down the lane in the scorching sun. When it was Nancy's turn, he would hardly move.

We all cracked up when she stood in the cart in a skirt, doing her Sieg Heils and Hitler salutes. Once she was seated, Grant handed her the reins and began pulling on the bridle to try to get the burro going, but he could only get him to walk. Grant came up with the idea of tying a rattle to his tail to make him move, so he put just enough gravel in a can to get a good rattle going and tied it on his tail. He prodded the burro to start him out, and when the burro heard the rattle in the can, he started to pick up his pace; he even began to trot. The faster he ran, the louder the can rattled. By the time he was a block from the house, he was bucking and kicking, trying to get away from that loud noise coming from his tail. All this time, Nancy, looking like the terrified Furher, was pulling at the reins, trying to get him under control; and Grant was running after them as fast as he could, yelling at the burro, trying to catch up to them, grab the harness and calm the beast down. Finally, the burro, having lost all control with all the rattle and commotion, circled back, running all over the lane, bucking and kicking trying to break free. He ran through an old, broken-down gate that had big, prickly pear cactus on both sides of it. Der Fuehrer was holding on for dear life. At last, the exhausted burro kicked free of the can and slowed down enough for Grant to catch up to them and grab the reins. After much coaxing and calming, the poor burro came to a stop in the backyard. Nancy, Der Female Fuehrer had just finished the wildest ride of her life. Her face was ghostly white and her moustache was smeared all across her upper lip. All of us kids stood and laughed until our sides hurt from watching up the comical scene, as Nancy awkwardly climbed down from the cart. We never thought once of her being in danger. We were too busy watching Adolph Hitler in a skirt with long hair bouncing along in that old, two-wheeled cart behind a kicking, bucking burro.

♪

One hot afternoon Uncle Grant and Aunt Minnie brought their kids out from Bisbee and left them to stay with us for a week. The day after they left, young Grant was walking up the road in front of the house as Mr. Harsha pulled up with his pickup full of slop, almost hitting him, as he swung wide to pull up to the gate.

Grant junior made a smart remark to him as Harsha got out to open the gate, an argument started, Grant could see that he was drunk and started walking away. When Harsha got back to his truck, he grabbed a .22 rifle from the rack behind the seat and yelled at him to start dancing. We heard several shots as he started firing at the ground near Grant's feet.

Terrified, he ran to the house as fast as he could, al-most falling as he ran up the steps of the porch screaming, "Aunt Lil, Aunt Lil."

The rest of us, hearing the shots and all the commotion, came running from the back of the house. We got there just in time to see Mr. Harsha laughing, walking back to his truck with the rifle in his hand.

Mother had us all go in the house, and after Mr. Harsha got busy feeding his hogs, Mother went over and told his wife about him shooting around Grant's feet. Dad went over and talked to them both the next day after he came home from work and made it quite clear that Mr. Harsha's little "bullet dance" better never happen again.

His wife knew he was an alcoholic and would hide his bottles of wine while he was feeding the hogs.

He never gave us trouble after he and Dad had their talk. He never came in our yard unless he was looking for one of his animals, which seldom got out and came into the yard. When they did, after the incident with Grant, he was always polite and apologetic.

The Harshas had a good-sized packing shed for their produce. They had two big vats of cold water for washing the vegetables when they were trucked in from the fields. The radishes, carrots, and onions were tied in bunches before they arrived at the packing shed. The two three-foot deep vats that they were washed in, one for washing and the other for rinsing, were always running with fresh water, and the over-flow ran into a pipe that ran into an irrigation ditch running along the side of the road.

Some nights after the trucks were loaded and all the workers had gone, with the dim night-light shining in the shed, Dave and I would sneak over and play in the concrete rinse vat and cool off. When Tom, Dave and I weren't playing with our boats and submarines in the shallow ditch, we waded in it and caught bullfrogs. It was a cool and fun place on those hot summer nights.

We got two more dogs while we were living at the Harsha place. A long-haired, yellow dog we named Abbey; and my dog, a puppy, a short-haired dog that Lila named Waddles. He was a mix between a dachshund and god-only-knows what else. His coat was gray with black spots like a Blue Heeler, and he had real small black spots like freckles on his nose and stomach. He was a mix. He was a loving dog, a dog the whole family liked; but he ended up being one of the ugliest dogs we ever had.

He would come into the barn at feeding time and gorge out on laying mash or the oat mix that Mother fed the cow at milking time. He always had plenty to eat when we fed the other dogs. We never knew how in the world he ever acquired a taste for cow feed. I guess he liked the taste of the sweet oats mixed in the feed. He would eat till his stomach looked like it was about to pop. When he left the barn, it was so big and bloated he waddled when he walked, that big stomach almost dragging the ground. Lila, the first one to see this, appropriately named him Waddles.

♪

Bill got his first leave while we were living at the Harsha place. He didn't have to report back to his ship for thirty days. We were all so happy to see him, we were all so proud of him. We all ganged around him asking all kinds of questions. He had seen a lot of action in the Pacific, and had traveled to a lot of different places.

He brought home souvenirs from Calcutta and the Fiji Islands. He brought Mother a beautiful picture of the Taj Mahal from India—a famous mausoleum built of white marble in 1631 by the Emperor Shah Jehan. The mausoleum was built for the tombs of his favorite wife and himself.

The picture was a mini-tapestry sewn with silver thread on dark blue velvet that Mother had put in a frame. He also brought back a beautiful curved dagger, a Gurkha knife shaped like a boomerang and a hand-carved outrigger canoe from Tonga, a group of islands in Fiji.

When he first came home, he practically jumped out of his chair every time the screen door slammed, which was quite often with us kids running in and out.

He had already served on two ships. The first one named the M.V. (Merchant Victory) Day Star was torpedoed and sunk; luckily, there were no lives lost.

He and the other Gunners Mates almost went down with the ship.

It was about two in the morning when the torpedo hit. The explosion knocked just about everyone not on watch out of his bunk. Bill scrambled to his feet and ran topside to his battle station to man one of the 40 mm guns at the stern of the ship. He and the other Gunners' Mates manned the guns as the rest of the crew was lowered over the side in lifeboats. The Gunners' Mates were the last ones to leave the ship. They hoped to get a shot at the submarine if it surfaced.

Most of the lifeboats pulled as far away from the sinking ship as they could to keep from being sucked under when the ship started its fatal plunge, leaving one life boat to go back and pick up the gun crew.

The ship was on its way to the bottom as they were pulled into the last lifeboat, narrowly missing being pulled under as the ship went down.

The second merchant ship was the one he had taken leave from. It had been under attack by Kamikaze planes as the ship steamed toward Pearl Harbor on its way back to the States. He and the rest of the crew were pretty shaken up, but luckily, again, the ship wasn't hit nor any crewmembers lost.

One day while Bill was home on leave I was playing near a small irrigation pipe that ran under the road not far from the house when Tippy ran over to the pipe and began to bark. I left him barking and ran back to house and told Bill that there was a small animal, probably a bunny rabbit, in the pipe and asked him if he would help me get it out. He grabbed a flashlight and went back with me to check it out. All he could see were its eyes when he shined the light in.

He went back to the house and made a snare like a shepherd's crook. It was all I could do to hold the little dog as he slowly dragged the "bunny" out of the irrigation pipe. To our surprise the so-called bunny was a good-sized skunk and just as it came out, it turned and sprayed Bill's legs and ran into the field across the road. Bill was lucky it just sprayed his legs. After four days of scrubbing down with tomato juice and using lots of shaving lotion, the odor finally went away. He thought the smell was also gone from the thongs he was wearing; he had scrubbed them several times and hung them on the clothesline.

A couple of weeks later when he was back aboard ship, several of his shipmates said they couldn't be-lieve they could smell a skunk that far from land. The dampness brought back the odor in the shower shoes Bill was wearing. He said that when no one was look-ing, he had thrown them over the side.

Tippy and the Rattle Snake

One late afternoon Dave and I were sent out to drive the cow back into the barn to be milked. It was hot and sultry; there had been hard thundershowers earlier in the day and there were still puddles in the cow tracks and some of the low places we walked down the cow path. We were always on the lookout for scorpions, centipedes, and most of all, rattlesnakes. Our tennis shoes weren't much protection.

Tippy had run ahead and was checking out the path when suddenly he began to bark. We knew by the sound of his bark that it was something pretty serious.

When we caught up to him, we could see he was barking at a large rattlesnake coiled up under a mesquite bush. He would run up just out of the snake's deadly strike and try to bite the snake as it slithered back under to coil up and strike again.

This went on for several minutes before we found enough big rocks to kill it. We were sure glad the dog or neither one of us got bit. It didn't take us long to round up the cow and head back. We searched every bush and every rock on the way. We had been warned about rattlesnakes, and that was the first one we had ever killed, one we had come up on without a grown-up around.

♪

By the mid-1940s, everyone was affected by the war and nearly everyone we knew had a member of the family or a relative in the service. When we went to the movies, the latest news was shown before the cartoons or the main feature came on, keeping us posted on how America was progressing in the war.

We were living within twenty miles of the Douglas Army Air Base so we got to see our share of war planes; sometimes as many as forty would fly over the school, squadrons of B-25s and B-17s with fighter escorts.

If we were in the schoolyard, we could hear the roar of the big radial engines as they approached. We would jump up and down waving our hands as the roaring planes passed over.

One afternoon at recess we all looked up as several squadrons of bombers flew over the school. All at once one of the fighters broke away from his squad-ron and circled back over the schoolyard, cutting the power as he flew low, low enough for us to see him in the cockpit. We jumped up and down yelling and waving. After passing he wiggled his wings and roared off at a steep angle to join his squadron.

One of the kids in my class was a real artist. He was always drawing when he had any free time, and he drew page after page of war scenes. He drew scenes of American and British planes in dogfights with German, Japanese, and Italian planes. He drew bombing scenes, and was exceptionally good at drawing German Stuka dive-bombers in action, the single engine Stukas almost in a vertical position when they released their bombs.

About two or three times a month, we would go to Douglas to do our shopping. After the grocery shopping, we usually went to Woolworth's 5-and-10 cent store.

Back in those days, there were no such things as shopping malls. The stores were quite different. We were fascinated by the way the clerks made change in some of the department stores. In the bigger stores, the clerks in the different departments would not ring up your sales and give you back change from a cash register. The clerk would make out a sales slip writing down the amount of the purchase and place it along with your money into a little round capsule that was attached to a cord, which ran up a line to the manager or cashier's desk.

The receipt and the change were returned the same way. Brother Tom, my cousin Dave, and I were always fascinated at how fast the capsule moved as it shot up to the office and down again bringing our change.

One Saturday we went to an afternoon movie and did our shopping after the movie. When we went to Woolworth's five-and-dime, we nearly always came out with a toy of some kind. This time Dave and I each bought a little plastic plane, and Tom bought a toy drum. The planes were neat little B-17s in fine detail. We enjoyed flying them all the way home, one of us on each side of the car sitting near the rolled down windows, watching the propellers turning in the wind while Tom banged on his drum.

A Brutal Beating

The worst whipping Dad ever gave me was when we were living at the Harsha Place.

It all started one Saturday afternoon when we were working in the barn and Dave and I got into an argument and Dad told us both to shut up. Dave did what he was told, I just had to have the last word and Dad reminded me that I was told to shut up. I was pretty worked up and sassed him back. I can't remember exactly what I said that caused it, but I well remember the hard whipping. He yelled at me and said, "Go to the house, young man, and I'll tend to you."

I'll never forget those words. I have never forgotten that day.

We both started up the road and when we were a few feet from the house, I took off on a dead run. I ran behind the house and started across our big open field.

My running from him made him furious. We never showed disrespect by running from our parents, especially when we were being corrected.

When he saw me take off across the field, he walked as fast as he could and climbed into our brown '35 Ford sedan, and slowly started chasing me. After being crippled up from polio, there was no way he could catch me on foot.

I ran in circles and zigzagged trying to get away. Finally after about ten minutes into the chase I was out of breath and stumbled and fell to the ground, and just as I was getting to my feet, his car was right along side of me. He got out of the car as fast as he could, grabbed hold of my arm, pulled off his belt and began giving me the whipping of my life.

I fell to the ground, and he was really laying it on when Mother came running across the field and stopped him.

I ended up with skinned places and big welts and bruises all over my ribs and back where the belt and belt had buckle hit. I think he would have hurt me pretty bad and put me in the hospital if Mother hadn't stopped him.

The Murder of Ed Miller

Not far from us lived the Smith family. They worked for the Harshas and lived in a run-down ranch house that they were renting from them. There were four in the family—Mr. Smith and his wife, stepdaughter Mary Frances, and a nephew who was in his mid-thirties. The older Smith was a big burly man, about six-feet tall, weighing over two hundred pounds.

The younger Smith was about as tall but not quite as big and heavy. The Smiths had a tenant living on the property who also worked for the Harshas. His name was Ed Miller. He was a tall thin man about fifty years old who paid room and board to live in one of the small buildings close to the house.

Ed too had been stricken with polio in his younger years. It didn't leave him crippled, but it did permanent damage to his neck muscles, causing his head to be drawn down close to his right shoulder. We could always recognize him from a distance even with his back turned because of this handicap. He drove tractors and helped in harvesting the vegetables and did other chores that needed to be done around the farm.

Mary Frances, stepdaughter of the Smiths, went to school at Double Adobe, but was in a different grade than Dave or I and we never paid much attention to her. We really didn't know her. There was quite a bit of distance between our houses and we couldn't see their ranch through the mesquite trees, so we had no idea of what went on in or around the Smith place.

One day when we were working in the fields, we overheard someone saying that Mr. Miller was saving money to go back to see his family in Illinois. He had no car and depended on the Smiths for transportation.

He paid for their gas when they took him to shop, go to the doctor, or to the barber.

When we started back to school the fall after we moved in, we never saw Ed again. We never thought anything about it because after the main crops were harvested, the Harshas always cut back on their work force. We thought Mr. Miller had probably gone back east to see his family.

One evening at the supper table, we found out what happened to him.

Dad was on his coffee break at work when he picked up the newspaper. There on the front page was a picture and an article about a body that was discovered in a dry wash near a bridge between Elfrida and Douglas. A soldier and his girlfriend had come upon his body when they were out hiking. The police were asking for anyone who might know the dead man to come to the morgue and identify the body.

After looking at the picture for a few minutes, Dad recognized him by the way his head was drawn down to his shoulder and the way he was dressed. It was Ed Miller. After work, he went to the morgue and identified him. He had been struck in the back of the head and his throat had been cut from ear to ear.

Three months later, the Smiths were arrested. The investigators had found bloodstains on the back and front seats of their car and bloodstains and hair on a claw hammer in the trunk. The Smiths were taking Mr. Miller to Douglas to the train station when they did the killing.

On the two-lane road between the Smiths house and Douglas there was little traffic, and there were stretches where you could see for miles ahead or behind you. There were few houses close to the highway and most of the scattered ranches were quite a ways from the road. There was hardly anything but desert, dry washes, a few bridges and mesquite.

After the police had done a thorough investigation, they surmised that the murder had been carefully planned. They said that Smith's nephew rode in the back seat while Mr. Miller and his uncle rode in front. The nephew picked up the claw hammer from the floor and struck Miller several times from behind, cracking his skull. They then pulled off the highway and finished him off. While the nephew held him, the older Smith used a butcher knife to slit Ed's throat from ear to ear, almost cutting off his head. They then drove to a suitable spot not far off the highway stripped him of anything of value, taking the $1200 they knew he had in the money belt he was wearing, and dumped his body.

Our Make-Believe Pickup

We never got bored living on our little farm. If we weren't doing our chores, feeding what little livestock we had or gathering eggs, you would find us playing in the backyard or in the irrigation ditch that ran along side the road. We played for hours with our cap guns, homemade bows and arrows and swords.

Sometimes we played with our make-believe pickup– that was, of course, if we could find the right-sized cardboard box to make the cab, one that would fit in the wheelbarrow. We cut doors and a windshield in the box and fastened the homemade cardboard dash, gauges drawn on it with a pencil.

The pickup was driven by Tom, the only one of us small enough to fit in box. He was the driver and we had a bumpy, and many times wild, ride.

One of my yard toys was my make-believe motorcycle. I used a big iron wheelbarrow wheel with a short pipe run through the center as an axel. On any given day, you could see me running around the backyard, bent over the wheel with my hands on each side of the wheel, gripping the axel as the wheel turned, barely missing my chin.

I know my motorcycle rides were pretty comical. Bill cracked up laughing when he saw me running around the yard with my head down and back end sticking up. We talked and laughed about it years later.

Not Big Enough for A Ball Team

The little school in Double Adobe School wasn't big enough to have ball teams to compete with other schools in or near Douglas and Bisbee. It had no gym and didn't have enough students in the different grades to make up ball teams. There was no ball prac-tice after school and most of the kids like us had chores to do when we got home.

The first thing we did was change clothes and help Mother feed the animals, including chickens, and then do our homework. By the time we finished that, it was usually too dark to play outside, so we either listened to stories on the radio, read comic books, or read Big Little books.

Big Little books were quite popular; I think every kid had one or two with his collection of comic books. They were little books about four inches square and two inches thick with a hard cover. They had a special feature we all enjoyed, showing the characters in the stories in action. As you quickly flipped the pages the characters in a little square in the upper right-hand corner came alive. On each page the character was in a slightly different position as was shown on the page before, which made him look like he was moving. It was like watching a miniature movie on the corner of the page.

Kids collected comic books and Big Little books and traded them amongst each other for ones they hadn't read. We did our trading on the bus on the way home or on the way to school. One boy in particular seemed to always have the latest comics of Superman, Batman, the Green Hornet, Black Hawk and Captain Marvel.

He lived near the beacon light in Elfrida, not far from McNeal. We were told his parents, the Millers (no relation to Ed Miller), did the maintenance on the beacon at the crossroads. One day when he got off the bus, Dave and I noticed a stocky Indian lady with a face like Geronimo, wearing homemade sandals, made specifically for her big wide feet. The sandals were made from sections of tire tread and held on with leather straps. She greeted him with a big smile as he handed her several Captain Marvel comic books. We later found out that she was his grandmother and Captain Marvel comics were her favorite reading material.

Fun With the Two Daves

Both my cousin Dave and my brother Dave were short. They were the shortest of the boys in both families. Being short, however, didn't interfere with their bravery. It never deterred them from taking risks, or doing something really dangerous like walking through a pasture of mean bulls when they were wearing, of all things, red or bright-colored shirts, just to take a short cut. If they weren't climbing as high as they could in a tree, or on the windmill tower, they were rappelling down a rope any place they could find a bank high enough.

Cousin Dave came out and stayed with us a lot; he even went to school with us for a while after he dropped out of the Catholic school in Bisbee. Once, he came out and spent a week with us and brought a miner's hardhat with him. We knew the perfect place to try it out.

In the neighbors' field among the mesquite bushes not far from the house was a good -sized hole where someone had started to dig a well or a hole for a big septic tank for a house, but for some reason, had abandoned the project. It was fenced off so no cattle could fall in, and one side had caved in making it easy for us to climb down to the bottom. We would take turns wearing the hat. Two of us would stay up near the edge and throw clods down hitting the hat, while the one wearing the hat pretended he was in a mine that was caving in.

After having the hat knocked off, and getting clobbered on the shoulders and back a few times, we could see that one of us might get hurt, so stopped and decided to play someplace else.

As we started back toward the house, my brother Dave stopped us, saying we should check out the orchard on the property next to piece we lived on. As we walked out between the mesquite, we could see the old orchard, or what was left of one, about a hundred feet or so on the other side of the fence.

After checking out the apple trees, and taking a few bites out of the dried-up, worm-eaten apples, we came upon what looked like remains of the foundation of a small house that had burned down some years before; and just a few feet away was a the tumbled down remains of an old outhouse.

We were about to head home when Dave Carroll, standing on some planks, yelled, "Hey you guys, come and see this." When we got there, he was dropping rocks down between the planks. He said, "Listen to this," and we heard the splash as the rocks landed in the water below. We were standing over a well.

I stepped off and the two Daves moved back the planks, letting in the sunshine. We could see a rickety wooden ladder hanging from bolts in a concrete ring that circled the opening of the well. As we started to throw in a few more rocks, we saw something moving on one of the ladder rungs. We kept splashing rocks getting it to move, so we could figure out what it was.

It turned out to be a baby owl, and of course, we had to have it. Cousin Dave and I held the ladder steady as Brother Dave went down and caught the little owl and managed to bring it out. I don't know what we would have done if the ladder had broken and he had fallen into the well. The well must have been twenty feet deep or more and we had no rope to throw him in case of trouble, and we were too far from the house to get help to get him out. At any rate, trying to get him out was a risky business.

JAMES P. CREEKMORE

3 CALIFORNIA

Dad never seemed to be satisfied living in any one place for any length of time. In most cases he moved to get a better job that paid higher wages; but sometimes though, I think he just liked to travel and see new places. Moving so much was sure tough on the family.

We would just get started in school and make new friends, and get acquainted with the neighbors, and get settled in; and the next thing we knew, we were packing up to move, to another town or to another state, like the migrant worker, like the "fruit tramp" that traveled from state to state following the harvest, never staying in one place. Now we were getting ready to move again, this time to the coast of California.

Vallejo

The application Dad sent in to the Navy Yard at Mare Island in Vallejo had been accepted. He would he work in a machine shop making parts for ships and submarines. His plan was to work there until the war was over and thought that by then, my folks would have enough money to buy another place, a place of their choice. We would also be able to see Bill when his ship returned to its homeport in Oakland, not far from Vallejo. According to the news, things were taking a turn for the better, our Marines were taking back the islands in the Pacific, and our U.S Army and Marines were beating back the Germans and Italians in Europe.

We had just settled onto our own little farm. Dad was working steady and both of the twins were working, and we had plenty of food on the table. Why move now, now that we had all this going for us? It was the best place we had since we left Kentucky (other than the place in Denver). Mother was in tears. It broke her heart, having to leave her nice vegetable garden, her chickens and animals.

She had saved $2000, money made from the vegetables she had grown on five acres. She sold all the animals, including the chickens, and we left, taking with us only two of the dogs, our little rat terrier, Tippy and Tom's dog, Abbey. We kids were in tears too when we were told we couldn't take Waddles. We had gotten attached to the ugly worthless dog and couldn't stand the thought of giving him up to one of the families in Double Adobe.

After getting rid of most of our furniture, we loaded down a large homemade box trailer with some furniture, kitchen utensils, clothes, and bedclothes, a full-sized mattress for Mother and Dad, and two small ones for us kids.

We took both cars, each with a water bag hanging out the window or tied onto something on the outside of the car. (A water bag was kind of a poor man's Thermos.

The canvas bags were filled with drinking water, and the bags were made so that the water inside seeped out through the bag just enough to keep it cold as the wind blew against it as it hung on the outside of the car.)

The twins and Lila loaded up whatever they could cram in the rumble seat of the twins' car, a 1931, seafoam green Ford Roadster, a car that Uncle Grant had helped them buy in Bisbee.

We led the way as we waved goodbye to the Harshas and pulled out of the yard with Dad driving the old '36 Olds, towing the big box trailer covered with a GI tarp, tied up in a bundle with the crisscrossed ropes holding it down.

Tom, Dave and I sat in the back seat holding the two dogs with our comic books and Big Little books on the seat beside us. We left before school was out that year, and we really hated leaving the school and our friends. I wasn't doing that good in school, but I still hated to leave. We all did.

Moving to California meant starting again in a big school, probably as big as Garden Home School in Denver, if not bigger.

We left before my eleventh birthday.

We didn't stay in motels but camped along the way like so many families of that time. Dad did all the driving. We didn't travel long distances at night because Dad was a careful driver didn't like taking the chance of getting in a wreck while driving in the glare of the oncoming headlights, especially from the big rigs. We always seemed to find a suitable campsite before dark. Back then there was a lot of wide-open space between towns.

Some of the towns of that time consisted of a grocery store, a filling station and maybe a post office, a church or a little park.

There were no freeways then, no tailgating in seventy-five and eighty mile an hour traffic. Driving across country was much easier, and much less complicated and stressful. You didn't drive along at eighty miles an hour to keep from getting rear ended while trying to figure out the confusing and sometimes complicated interchanges you faced when entering or leaving a good sized city. Back then we traveled at a much slower pace with far more courteous drivers.

There were a lot of people on the move at that time, many moving to be closer to defense plants. It was nothing to see car after car pulling trailers. People towed homemade trailers with vans, pickups, and cars. Many of the cars were like ours, made in the mid- to late thirties. We got use to seeing old, square-bodied, four-door cars full of kids and dogs with their mattresses, and sometimes furniture, tied to the top.

Sometimes we camped by old abandoned buildings for shelter from the wind. We tried to find safe places far enough off the highway to be away from the sounds and lights of the traffic. We kids always waited in the car while the grown-ups checked out the campsite. As soon as we got out of the car, some of us were sent out to gather firewood while the others gathered rocks to put around our campfire to keep the fire from spreading into an open field, and to set our skillets and coffee pots on.

We would pick out a nice flat place away from the fire, a place with not too many rocks, to spread a big canvas tarp and make our beds. Mother cooked our supper over the open fire. It was usually scrambled eggs, potatoes, and bread. We often had the same menu for our breakfast.

When we arrived at Vallejo, we found out there was a shortage of rental houses. Renting a house was close to impossible we were told; but, we could rent an apartment by getting our name on a waiting list. It was just a matter of time. We had no friends or family to stay with or help us out, so we ended up living in a tent right across the bay from Mare Island, between the San Francisco bay and Wilson street. Dad bought a GI tent that was big enough for us all to sleep in. It was made with an opening in the top of it to accom-modate small stovepipe for a wood-burning stove. There was room enough for a big bed for the folks and cots for the rest of us. Our clothes and other be-longings were kept in the trailer or piled in the corner of the tent.

Living in a tent was no real problem for us except when the rains set in. It was a mess, but having lived in some of the places in Colorado without running water, electric lights or indoor toilets, living in a tent for a few months no real challenge for us.

The twins weren't allowed to drive their Model A roadster in Vallejo. The folks thought it was too dangerous with all the traffic, especially with all the big transit buses ridden by both service men and civilians to get to different parts of the busy city.

Living near the ocean was a real experience for us. We had just come from one of the hottest, driest places in the country, a state full of cactus and mesquite bushes, with scorpions, tarantulas and rattlers, and now we were living in an entirely different climate with fog and rain, the bay not more than five hundred feet away.

We had never seen the ocean, much less lived that close to it. Everything was strange and new to us. When the fog came in, in the early mornings, we could just see the lights of Mare Island, and at low tide, the wind coming off the bay had that special ocean smell. We were close enough to hear the sounds of foghorns, whistles and bells coming from the water traffic, as it came and went from the island. All these things were quite exciting for us, the kids from the desert. We watched in awe as submarines returned from war patrols with bullet-riddled super structures and dented ballast tanks from depth charges. They made their way to the Navy yard, some limping in on their own power and some being towed by sea-going tugs.

The only submarines that Dave, Tom, or I had ever seen were the toys we played with in the irrigation ditches in Arizona.

An Attempted Rape

Work went on around the clock at Mare Island. Not only did Dad work, but the twins also got jobs on the island packaging submarine parts.

Vallejo was crowded with sailors and yard workers and their families. There were little makeshift plywood shops lining some of the streets going down toward the waterfront. There were dry cleaners, laundromats, tailor and uniform shops, and hamburger and hot dog stands. Gray government buses loaded with sailors and civilians alike crowded the streets. They were usually so full there was hardly standing room. Uptown there were penny arcades, restaurants, music stores and pawnshops.

Our tent sat a half block off Wilson Street, well grown up with brush and willows. The twins and

Lila usually walked to town taking the path that ran through the brush and willows up to Wilson and then on up to town.

One Saturday when the twins were on their way home, about half way between Wilson and the tent, they were startled by a noise coming from the brush behind them. As they turned to see where the noise had come from, Laura was grabbed from behind and spun around. By then, they could see what it was all about. Laura broke free as Leta pounded on the stranger, yelling at him to let Laura go. He stood in the pathway exposing himself, repeating over and over to Laura, "I'm going to rape you. I'm going to rape you." When Laura broke free, they both ran back to the tent. They got there, all out of breath, gasping out what had happened.

Dad grabbed his loaded deer rifle and headed up the path as fast as he could go.

Dad, with his slow gait, knew that he wouldn't be able to catch the guy if he had to run. His speed was a fast walk at best. He'd worked the night shift standing while operating a lathe. With his aching feet and crippled legs, running was out of the question; but, he was sure the attacker couldn't out run a bullet.

By the time he filled out the police report, the suspect was long gone. The twins and Lila were really cautious after that and always stayed close together in crowds.

♪

When we finally got our apartment, we were really happy, even if it was upstairs. It didn't take long for us to make new friends once we got settled. There were families in the apartments from states all across America. There was also a family from Russia, a woman, and her husband and young son who was about Tom's age and spoke broken English. We spent our weekends in and around the apartments, visiting or playing with the neighbor kids, or going to see the latest movies.

The twins and Lila were joking about stooping over when they put anything in the corner of a room. They had gotten so used to stooping to put anything in the corner of the tent that they couldn't get used to standing up straight.

We tried to keep the kids from bunching up near our apartment because Dad worked nights, and it was rough on him trying to sleep when there were kids around.

One day Lila noticed that the kids who were always near our apartment had disappeared. She asked me where they had gone, and I told her that I had told them that Dad had a slight mental problem and was very dangerous to be around, and that he flew into a rage if he didn't get his sleep. She cracked up.

A lot of the kids built models. Joe Yarris, one of the kids that lived a couple of doors down from us had quite a collection. We never refused when he invited us in to check out his latest stuff. He had tanks, trucks, ammunition carriers, half-tracks and all sorts of artillery pieces, all built to scale and detailed out with their proper markings, numbers and decals. Most of them were made of plastic and you could roll them into their fighting positions all decked out in their camouflage paint.

Joe Yarris not only had nice models of planes and tanks used in World War II, but he also was a real marksman with a bow and arrow. We were fascinated with his accuracy, his marksmanship. He practiced in one of the open fields that were on the many rolling hillsides above the apartments.

He would place a cardboard box a couple hundred feet away on the side of a hill and shoot the arrow up like he was aiming at something in the sky and when it came back down it usually hit the box dead on.

Since Dad worked the night shift, we spent most of our time playing outdoors as long as it wasn't raining. And, of course, when it was raining, we used our trailer, which was parked near the curb not far from the apartment, as our clubhouse. With the heavy GI tarp stretched over the top, we were able to play in it even when it was raining. It was low to the ground and easy to get in and out of with the tailgate down. We sat on the floor and played cards and checkers by candlelight using an orange crate for a table.

Mother was choosy about which neighborhood kids we let come into our little apartment. One kid in particular who liked to play cards with us was Richard Clogston. He was a tall, thin black-haired kid with a crew cut. His clothes always had a strong odor of tobacco.

Richard had lived in the Floyd Terrace project longer than most of the other kids, and seemed to know all the fun places to play. Dave and Lila had started sneaking off to smoke, and Richard, being a smoker too, always seemed to have cigarettes for the three of them. Mother and Dad weren't aware of them smoking at the time. Many people around us were smokers, and Dad was a heavy smoker, too.

Richard would join in our card and checker games and enjoy our pretend wine, which was a grape or strawberry soda.

One Sunday evening, he decided to show Dave and I his secret hiding place. We couldn't imagine where we were about to go. We followed him up the street behind the apartments. On one side of the street where there was no parking near the sidewalk, a manhole led down to the storm drain under the street.

After Richard made sure no one was watching, he quickly lifted the edge of the manhole cover with a metal hook and slid it back exposing the ladder going down into the big concrete storm drain below. He held his flashlight while Dave and I climbed down to the bottom; as we stepped off the last wrung into the big cement pipe, Richard climbed down and slid the lid back over the hole.

When we stepped off the ladder, we found that we couldn't stand up, but we could easily walk slightly stooped over. As Richard led the way with his flashlight, we noticed a little stream of water running downhill through the sandy bottom of the storm drain. We were able to step around it without getting our feet wet. The big pipe was pretty clean other than small sticks, parts of newspapers, and an occasional bottle or can.

We walked in about a hundred feet to what Richard called the "junction," a wide place where several drains from different streets came together, dumping their contents into the one big pipe.

He stopped in the middle where a small upside-down bucket had been placed on a sand bar and took a large candle from his jacket pocket. He had Dave hold the flashlight while he lit the candle. Dave and I stood gazing at the big candle as it brightened up the big storm drain.

It was a scary place for us as we looked at the dark concrete pipes branching off from the junction. We were wondering just what kind of creature might be living down there in one of those pipes that branched off. As Dave and Richard stood there smoking their cigarettes, I kept waiting for something to come charging out of one of the drains. I guess I had read too many scary comic books and heard too many spooky stories. We only went back a couple of times, and when the folks found out, it was all over for us.

♪

One of the sports we really enjoyed was kite flying. We would fly our store-bought kites until they were too battered up to get in the air, and then we would repair them. We took wooden cross pieces from the damaged kite and built new kites using newspapers and paper paste, and strips of old sheets, and used rags for the tails. The homemade ones worked just as good, and sometimes even better, than the store-bought ones.

With the twins and Dad all three working, we were able to have nice clothes and shoes for school, and even had money to pay for music lessons if one of us wanted learn to play an instrument. Of course, I chose one the hardest of all to play–the violin. The place where I took lessons also provided the violin to prac-tice on. I was even able to bring it home to play, what a treat.

I rode the bus back and forth from Floyd Terrace to the store downtown where I took my lessons. I did pretty well for the first two weeks. I could even saw out the song Don't Fence Me In, a popular tune of the time. I sort of played it like the teacher, only I played it by ear.

The third week, I was home practicing some difficult scales, over and over and over again, trying to get it right, sawing out the off-key notes. One evening when practicing, I flew mad, having one of my temper tantrums. I flung the violin at the bed, and missing the bed, it hit the wall, breaking the neck loose from the body.

Oh boy I was in for it; there was no way to explain it away. Mother gave me a hard whipping, and then had Dave help me glue it back together. After the glue dried, we left the strings loose when we put it back in the case hoping the glue would hold. Lucky for me the music teacher never opened the case when I set it on the counter, telling him it was my last lesson, and that it was too hard for me to learn to play. I politely thanked him, left the store and caught the bus back home. I could just imagine the surprise he was going to have when he tightened the strings to tune it if the glue didn't hold.

I felt bad after leaving the violin off at the music store because the people had treated me so nice and had been doing their best to help me learn how to properly play the nice instrument. Many nights thereafter, I regretted what I had done and knew I owed them an apology.

Going to school in Vallejo wasn't much fun. It seemed like I was getting in a fight just about every other day. I wore nice clothes, so they never made fun of what I wore. It was other things they said and did that made me mad, like intentionally hitting me and some of the other kids in head with a volley ball, or pushing us around on the schoolyard, or making fun of the way my hair was cut, telling me my head looked like a "grass hut"– the usual teasing, like kids do. After they found out that I had a short temper, they teased me all the more. I would get in fights after school with white kids as well as black kids.

We were all glad to know that the war would soon be over, and when it was, we would be leaving Vallejo.

When it was time to go, Lila and the twins didn't have to worry about leaving boyfriends behind. In fact, I can't remember them having any of their dates come to the house, except on a couple occasions when two guitar players came by and played for us; one of them was sweet on one of the twins.

They were both pretty good musicians, especially Huey, the steel guitar player. He played the latest hits and really impressed us when he played the Steel Guitar Rag, one of the popular tunes at the time.

Another time, Al, one of Leta's boyfriends, came by and took Tom and I fishing in the bay to show us how to fish. We went along see the ocean more than any-thing else. Al picked out a spot that was safe for us to fish and explore. Tom caught a couple of small fish and I ended up pretty sick with a stomachache and sore throat. We had never been around saltwater be-fore and while Al was showing Tom how to use a fishing reel, I was exploring some of the tidal pools about forty feet away. I was thirsty and made the mistake of bending down and taking a drink from one of the nice clear pools. The water looked clear and clean as the mountain streams we drank from in Colorado. I got down on my hands and knees and took a big gulp and came to my feet sputtering, coughing and throwing up. It was too late by then; I had swallowed enough to make me good and sick. Al and Tom came over to see what was wrong, and that was the end of our fishing trip.

I must have drunk a gallon of water when we got back to the apartment. I felt like an idiot when Al came over and asked what was wrong I was almost too embarrassed to tell him. I definitely learned that sea water was not for drinking.

We left Vallejo in early May of 1945. The folks were undecided as to whether to go north or south. The Carrolls had left Riverside, California and were now living in Burns, Oregon. Mother wanted to move to Florida where she could have another farm. To settle the argument, they decided to draw straws. Long straws meant north, short straws south. Mother made the first draw. The one that drew two of the same length in three tries would determine the direction we would go. After three tries she ended up with three short straws. That decided it. We were going south. The plan was to go to Florida.

Leaving Vallejo, leaving Floyd Terrace the apartment complex, was nothing like leaving our farm in Double Adobe. We were happy to know that we wouldn't be going to school in Vallejo in the fall. We were also happy at just the thought of going someplace where it wasn't so crowded, someplace where we would have some space, maybe even have a nice big house in the country and have farm animals again. Most of the kids we knew in Floyd Terrace were going to be leaving also. Some were going back to the state they had lived in before the war. Some were going to move to other parts of the state.

We loaded our clothes, beds, and mattresses and the few pieces of furniture we had accumulated along with most of our toys, leaving out a few out to play with along the way.

Bound for Florida

Dad checked the trailer spare bolted to the front of the trailer and after we finished putting water in our water cans, packing our cooking utensils and groceries. He closed the tailgate and lashed down the tarp. Then we started loading the homemade 1932 Chevy flatbed pickup, which Dad had talked the twins into trading for. They were pretty sad about trading their 1931 Model-A. They had just gotten use to their roadster and really hated to part with it. They had traded a nice-running, sporty, convertible roadster with a "rumble seat" for a cobbled together, homemade Chevy pickup.

The pickup had once been a four-door. The owners had cut the body off just behind the front seat and closed in the cab by bolting in tightly fitting boards with a small window in the center. The flat wooden bed with its sideboards and tailgate sat pretty high off the ground and wasn't the easiest thing to load.

After the tarp was put on and tied down, we all went back upstairs and made sure everything was clean. We said our goodbyes to the few friends we had, hooked up the trailer to our '36 Olds, and headed south.

We started out with Tom's dog Abbey lying on the floor, and Tippy lying comfortably in Lila's lap. Lila rode part of the time in the car with us, and part of the time with the twins. Tippy had been teased so much that he would growl and snap at the slightest push or tug. We all gave him plenty of space when we had him in the car. We didn't dare tease him. For one thing, we could get seriously bit by teasing him in such close quarters; for another, none of us could afford to upset Dad when he needed all his concentration on the road. Plus, if Dad had to stop, we knew we were in for a good whipping.

I think the trip between Vallejo and Texas was one of the most tiring trips we ever made with Tom, Dave, Lila and I crammed in the back of that old Olds with the two dogs, half the time spitting out dog hair, wondering when and where our next rest stop would be. Back then, they didn't have rest stops like the highways of today.

Traveling down Highway 99 towing the trailer, never going over fifty miles an hour, going forty-five a big part of time and being practically blown off the highway when big rigs passed was an experience all its own. Dad couldn't stand the glare of the lights from all the traffic after the sun went down, so again, we found a suitable spot before sunset, a spot far enough away from the traffic noise and glaring headlights, and camped for the night, heading out again the next morning after we had our breakfast.

We pulled off 99 several times for groceries and gas in some of the small farm towns along the way. It seemed like every little town had at least one orange juice stand, a little building that stood out as you came into town; they were little round buildings, shaped like and painted like oranges, with porch like counters. After topping off the gas tanks, we would stop at a grocery store for Mother to pick up stuff for sandwiches. We enjoyed the delicious sandwiches of bologna and longhorn cheese on rye bread, and a nice juicy dill pickle.

Dad always took spare parts with us on long trips– spare fuel pumps and water pumps and fan belts, and tools and patching for fixing flats, in case we broke down on the road. The little towns were few and far between, and you just never knew where or when you might have car trouble.

After leaving Bakersfield, California we kind of settled in for our long trip, just poking along across country, watching the big rigs go by and gazing off at orchards and little farms on both sides of the highway as we traveled farther and farther south.

While Dave, Lila, and I were talking about the different things we were seeing on and off the road, Tom was humming the first five notes of the "Steel Guitar Rag" over, and over, and over again. He hummed the tune for miles on end. Lila noticed Dad moving rest-lessly in his seat. She looked over at Dave and I and started to smile as she pointed to Dad as his head begin to get red. We buried our heads in our hands to keep from busting out laughing as Dad managed to turn half way around, scowling at Tom, saying, "Young man, can't you find some other song to sing?"

We had just gone through Riverside and had stopped outside of Perris for gas when Dad noticed that the trailer hitch was cracked.

We were lucky–just down the highway not far from the gas station was a welding shop. We were delayed a couple of hours while it was being welded. It was a good time for walking the dogs and having our afternoon meal of bologna sandwiches and water.

After the hitch was fixed, we headed east toward Blythe. The biggest challenge for our little caravan was Whitewater Grade, a fairly steep road winding up through the mountains between Palm Springs and Blythe.

We got into Blythe late in the evening and camped by the Colorado River. Mother got out the big skillet and fixed our supper, and we got out the mattresses and made our beds. We spent a big part of the next day in and around camp. It was a fun place for us kids. It was nice and warm and we were able to find a shallow safe place to play in the river and clean ourselves up.

After we all got cleaned up and everything packed once again, we stopped at a gas station, filled our water cans and gas tanks, bought a couple of canvas water bags to cool our drinking water. The bags were designed to be hung from rearview mirror braces or from the door handles of cars and trucks made in the 30s and 40s. The bags did not drip water but were always damp after being filled. The water was cooled by the wind, which made it desirable to drink for both people and pets while traveling the blistering highways on long trips. We were headed east.

At that time, in the forties, before there were freeways, gas stations could be found in some pretty remote places, especially on Route 66. There wasn't much in the way of scenery in the long stretches of highways that crossed the states of Arizona, New Mexico, Texas, and Oklahoma. Taking long trips across most of this country was pretty boring to say the least. And, it could be troublesome and even dangerous if you took the wrong turn and drove out some remote area and found out you didn't have enough gas to get back to the main highway. There wasn't a rest stop every so many miles, or call boxes or cell phones to contact someone if you were having trouble. If we had to go to the toilet, we pulled off the road and went behind a bush or under a bridge.

We broke up our boring rides by playing cards, counting railroad cars if we passed a train puffing up a long hill, or reading Burma Shave signs that were spread out between towns.

Burma Shave signs were clever little signs, sometimes spread out over a mile, that were poems advertising Burma Shave, a popular shaving cream. Each sign had just a few words of the poem. I remember one series of signs in particular that went like this:

> *To kiss a mug when it's like cactus*
>
> *Takes more nerve than it does practice*
>
> *Burma Shave*

None of us ever forgot the delicious Pepsis and tall Nehi sodas we enjoyed when we stopped for gas. We didn't get them very often, but when we did they were a real treat. The Nehis came in strawberry, orange, crème and grape. There wasn't any such thing as pop-tops back then; you used a bottle opener. They were so charged with carbonated water that they literally smoked when they were opened.

The bottles were clear glass and there was nothing prettier than a big Nehi soda, especially after drinking lukewarm water from a water bag while traveling in a hot car. We drank our sodas very slow, savoring every drop.

The cars in the thirties and forties had no air conditioners; the inside was "cooled" by the hot air blown in through the rolled down windows. Just outside a small town in Arizona, we were flagged down by a couple of Indians who walked right out into the middle of the road.

They were mean-looking Apaches, and there were several more on the side of the road near a stand selling beautifully hand woven blankets, some trinkets, and some turquoise jewelry.

By the time we came to a stop, several had surrounded the car and the old Chevy pickup behind us that the twins were riding in. They came right up to the car almost sticking their heads in the rolled down windows, and some were trying to lift up the tarps on the trailer and pick-up to see what we were hauling.

With the calamity of both dogs barking at the tops of their lungs, Dad convinced them we weren't going to buy anything, and we went on down the road and found a place to pull off and camp. We were about three hundred miles from El Paso.

The next day we got as far as a the little town of Canutillo, Texas, about twenty miles north of El Paso, when we started having major problems with the Olds. Between Anthony, New Mexico and Canutillo, the engine began to knock, and Dad knew we had a bearing going out. We had to stop soon to keep from completely wiping out the engine. When we pulled into Canutillo, it was hammering pretty loud.

We pulled into a filling station across from Lawrence's garage and grocery store. Dad went across and talked to Glen Lawrence at the garage, and found out he was the oldest son of the man that owned both the store and the garage. Mr. and Mrs. Lawrence, both in their late sixties, were a real nice couple that had been there for years. They were known by most of the cotton farmers in the valley. Dad explained our predicament, that we were low on money, on our way to Florida, and we needed a place to stay until the car was repaired.

When Glen found out that Dad was a machinist as well as a mechanic, he agreed to let him work out the bill at the garage and told him of a vacant house on the other side of the river that we might be able to rent pretty cheap. The old adobe house had electricity, but didn't have an indoor toilet or running water.

We rented it. At the time it was about all we could afford—the old adobe house on the west side of the Rio Grande, set close to a paved road that threaded its way south between the huge cotton fields.

It sat right in the bend of an S-curve, and at sometime long before we moved in, someone had lost control of their car or truck and crashed into the side of the house with such force that it pushed in some of the adobe bricks and cracked the slab floor. The crack left one part of the slab higher than the other and made it miserable walking from one of the bedrooms into the living room. We were always stubbing our toes on the uneven floor.

The house was plenty cool both summer and winter. There were no lawns in either the front or back yard, and there were no sidewalks. We made our own sidewalk in the back by placing a few flat rocks and a couple of short planks that led to the back door.

The place was a real mess when it rained. The ground around the house was adobe and when you stepped off the makeshift sidewalk, the heavy adobe would stick to your shoes like putty. It was tough to scrape it off before going in the house.

There was no running water in the house, but it did have a well with an old pitcher pump in the backyard. The well was near some salt cedar trees.

The pump worked fine and there was plenty of water, but it was what we called "alkali" water, too bitter to drink or use for cooking. The water that splashed to the ground around the pump looked like a winter frost. We were told that the salt cedar trees caused the water to be the way it was.

Dad rigged up the box trailer for hauling our drinking water and water for cooking by removing the sideboards and tailgate, and making braces that cradled a big wine barrel with a bunghole on top, through which the keg or barrel is filled. It had a big wooden plug, a bung, and a faucet on the back end where we filled our buckets and jars. He hauled the water from the Richardson Ranch. Mr. Richardson was our landlord.

Dad worked at Lawrence's Garage just long enough to pay for the repairs on the Olds and keep up payment on our grocery bill at the store. He then took a job in El Paso working as a machinist and mechanic at a military bus garage. The twins were also working in El Paso for Western Union. Dave and Lila went El Paso high school, while Tom and I attended the Lone Star Elementary.

We hadn't been living there long before Bill, after serving his four years, got discharged from the Navy and came home. He didn't want a job picking cotton, so he took a job about forty miles away in Las Cruses, tending bar. He also played his saxophone and guitar there on Saturday nights for extra money.

Later on, near the end of the summer, Dad was of-fered a job on Guam in the Marianas Islands in the Pacific; the job paid a lot more than he was making in El Paso. He talked it over with Mother and they de-cided he would work there a year and come back, and we would get the necessary shots and all and go back with him, or we would go to Florida as was first planned.

We were pretty much settled in when he left. We were none too happy to be living in an old broken down house, but we didn't have much choice.

After we were there a couple of months, Mother bought a couple of laying hens and an old, white Leghorn rooster. She had them penned up under the shade of the salt cedar trees near the pump. She had, in the meantime, started working part-time at the Lone Star Café in Canutillo and she soon found that the chickens were more trouble than they were worth. She made chicken and dumplings out of the hens, but kept the old rooster that was too tough to eat.

The pen was left open and the rooster was left to roam. He hung out around the pump where we dropped scraps of food and would sometimes greet us at the back door. At first, we just kicked at him and he would jump out of the way. After a while, he would charge you as you walked away, hitting the back of your legs with his sharp spurs, sometimes drawing blood.

You never knew when he might attack. When we walked into the backyard coming in from school, we usually grabbed up a stick or something to fend him off. If you came out the back door headed for the outside toilet, you would hear his mud-caked feet pounding across the hard adobe yard. It made a clubbing sound as his big toes, caked with dried adobe, pounded across the yard.

Not long before we left Canutillo, we gave him to the Wells family, a big family of several kids that lived in a small stucco house in the middle of a cotton field about a half-mile away. Bill said that with his age and muscular legs and thighs, it might take a week to cook him, and we'd be lucky if we could even stick a fork in the gravy.

I think the Lone Star Elementary in Canutillo was just about the most disciplined school that Tom and I had ever attended. We even hated it worse than the one in

Vallejo. The little school that sat up on a hill was made up mostly of Mexican farm workers' kids and some pretty well-to-do ranch owners' kids.

Some of the Mexican kids were quite old for the grade they were in. One of the things you could expect in the Arizona and Texas schools was that if you didn't make passing grades, you would be held behind in the same grade the next fall unless you were tutored over the summer to bring your grades up. There were both white and Mexican kids who were too old for the grade that they were in. In both Arizona and Texas, if you failed the grade you were in, you had to take the same grade over the following year.

One student in my fifth grade class who I will never forget was eighteen-year-old Rodolpho Gollegos. He was the oldest pupil. He was too pre-occupied with the pretty girls or something going on outside to pay much attention to what was going on in the way of schoolwork. He was once caught playing with himself in the middle of a geography lesson.

Rodolpho wore bib overalls and always sat in one of the back rows holding the big geography in front of him. He held the big book with one hand, while using the other to play with his penis while looking at pretty Beatrice Carosco or Herlinda Flores or one of the other pretty girls up near the front of the room.

This went on for several weeks until one of the girls that came in late and had to take a seat in the back row caught him in the act and told the teacher. After a good paddling and a conference with his parents, he was taken out of school and suspended for a month. The school had some pretty strict rules and the teachers definitely believed in enforcing them.

Every teacher had his or her own paddle. I had been paddled in Lowell and Double Adobe, but it was nothing like being paddled at Lone Star. It seemed like I was always in a fight over something. It was every bit as bad, if not worse, than the school in Vallejo. If it wasn't in the classroom while the teacher was out, it was on the school playground. White kids as well as Mexican kids seemed to always find ways of making fun of or bullying the new kid. If they weren't snickering and making fun of your clothes and shoes, they were tripping you or shoving you while in line at the drinking fountain. And like the other schools I had attended, the bigger kids had their fun bouncing volleyball or basketball off your head when you least expected it and the teacher wasn't watching.

I had quite a temper and Tom and I had no big brother to turn to help us out when we got in trouble. I was always coming home with my nose and elbows skinned up. Mrs. Skinner, my class teacher; never paddled me but I was sent to the principal's office for slapping a Mexican girl and pulling her hair. She had shoved me and spit in my face when I got on to her for crowding in line ahead of me at the drinking foun-tain.

The kids all said that Mrs. Rugely, the principal, gave harder whippings than any of the teachers in school. It didn't take but a few minutes to convince me that what they had said was very true.

I was ordered to put anything I had in my pockets on her desk and pull my pant legs tight against my legs so there would be no air space or cushion between the pant legs and skin. I was then ordered to bend slightly over her desk. She then took the big paddle she had laying on the desk and started in. By the time she was through, my butt and legs were on fire. Never in any of the schools that I had attended, had I been paddled so hard. After that hard paddling, you can believe I never did anything bad enough to be sent to the principal again.

♪

The twins had gone to El Paso to get jobs. They were working there a couple of weeks before they started dating ex-GIs, veterans that had seen action in the war.

Johnny McCoy, Laura's boyfriend, was from Canutillo. He was helping his mother run the family cotton farm.

Leta's boyfriend G.C. Hughes was also a Texan, from Colorado City.

We didn't know at the time that G.C. had been awarded two Bronze Stars for his bravery while fighting in the Pacific. Johnny had also put in his tour of duty in the Pacific. Both men had served in the Army.

Laura was married in May 1946 and shortly thereafter leased and ran the Lone Star Café in Canutillo. When she became pregnant with their first child Michael Patrick, it eventually became too much for her. They gave up the café, which was then leased to Mr. and Mrs. H.E Baker. In March of 1947, Leta married G.C.

After Laura and John turned over the restaurant to the Bakers, Johnny leased the Chevron service station there in town renaming it Johnny's Chevron.

He did very well servicing the trucks, pickups and cars, changing tires and servicing the various pieces of machinery for the many farmers he knew in the valley. Many times when you passed Johnny's Chevron service station, you would see him working on an engine or changing a big tractor tire. He and Laura bought a home at the north end of town a short time before her son Michael Patrick was born.

Mother worked for a while at the laundry and then went to work for the Bakers at the Lone Star Café.

The Bakers also had children that attended the Lone Star Elementary. Tom and I would never forget the Baker boys, Jackie and Jerry, Jerry was the oldest. They both had full-sized bicycles with no fenders or chain guards that they rode to school, and on the weekends you would see them pedaling down the river levee with their right pant legs rolled up to keep them from getting tangled up in the sprockets. Their fishing poles and the small lard buckets they used as bait-and-tackle cans were tied to the handlebars.

Those buckets usually held sinkers and hooks and a couple of bacon rinds tied to a piece of fishing line that was wrapped around a small stick used for fishing for crawfish. They sometimes stopped by the house and showed us their catch. Sometimes the buckets would be full of crawfish.

They said that the best crawdad fishing was behind a chili cannery near Canutillo where the warm water from the cannery dumped into the irrigation ditch.

Tom was only eight and he and I were not allowed to play in the river without the presence of a grownup; at that time neither of us knew how to swim.

When we did get the privilege, we enjoyed playing in the cool water and catching bullfrogs in the ponds near the river and the small soft-shelled turtles that were buried in the sand. They were easy to spot. All we had to do was look for the two little holes in the sand where their long noses stuck out.

Tom and I also played in our dugout. The dugout was our little hideaway. We worked a couple weeks digging the hole for it in the backyard. The digging wasn't so hard after we dug down through the first couple of layers of adobe.

When it was finished the little room was about five feet deep and about five-foot square. We used an old rear tire off a John Deere tractor that had been lying at the edge of the cotton field to help support the roof. We laid it flat over the hole, and covered it and the short entrance tunnel with scrap lumber, some from the backyard and some that we dragged home from the foot of the levee. We covered it with dirt, leaving a slight hump in ground in the backyard.

The cramped little room was just big enough for Tom and I and sometimes Tippy. We only enjoyed it for about three months, until the first hard rain flooded it and we had to give it up. After it dried enough for us to work on it, Mother made us backfill it, and roll the old tractor tire back to the edge of the cotton field where we had found it, and stack the boards we had used for the roof, which we had shoveled dirt over to cover up the worn out tractor tire.

♪

I wasn't the only one getting in trouble while going to Lone Star.

The Baker boys, Jackie and Jerry, had their share of fights and paddlings too. The Baker boys had rock fights after school with the Mexican kids. I guess they got their accuracy from "chunkin" rocks from the levee at things floating down the river.

Although we never saw it happen, we knew their aim was so good either one could have easily clobbered you in the head from forty feet away. Jerry wore glasses, pretty thick ones at that, and we often wondered how he could do it. They were both good with sling-shots, too. They were always telling us, using their Texas slang, of their fights with "Messcans" and "chunkin" rocks at them.

Several times as we walked down the hill after school, we watched the Baker Boys jump off their bikes and "chunk" rocks at the Mexicans, then jump back on and charge down the hill.

One weekend they rode over on their bikes to show us their latest homemade weapon. They had made a pipe gun and the four of us went down near the levee where there was no one around and tested it out. It was made out of a water pipe about eight inches long and as big around as a quarter with an elbow on one end, and a plug in the other end as a short handle or pistol grip giving it the look of a pistol. They had drilled a hole on top of the elbow for a fuse to stick out. The fuse was lit with a wooden match from the box one of them carried in his pocket.

Loading it was pretty time consuming. It took several tries getting the fuse just right to come out the hole for ignition. A firecracker was dropped down the barrel fuse first so the fuse came out the hole. Next went a wad of paper against the firecracker and some sand and small gravel. Jackie tried it first. He lit the fuse and stood clear of the rest of us and aimed toward the levee. There was a muffled sound as the firecracker went off, blowing the sand and gravel from the barrel. Jerry ended up burning his hand between his thumb and forefinger where he had gripped it too close to the fuse. That didn't stop him from doing it again! They had several firecrackers and fired it several times. Tom and I had been around rifles and handguns all our lives but never asked to shoot it. Fooling with a homemade pipe gun was pretty dangerous; it's a wonder no one was seriously hurt.

Neither Tom nor I participated much in any sports. At recess we sometimes played volleyball and baseball, but never stayed after school to practice with a team. The only thing we stayed after school for was when we had been disciplined for talking in class or some other minor offense. We normally left the school grounds as soon as school let out.

Sometimes the twins were waiting for us in the old homemade pickup, but most of the time we walked home.

We usually stopped at the Lone Star Café if Mother was working. The days she wasn't working, we would stop at Buck Brisco's Texaco station and get a nice cold soda, either a Nehi orange or strawberry or a Pepsi.

After finishing our sodas we made our way up the street to Lawrence's store where we picked up the material for the next day's lunches and a Pepsi each for of us to have on the way home. We had a running tab that Dad paid at the end of the month.

Mr. Lawrence worked at the counter. He was well over six feet tall, had a big moustache, and wore a big western hat and cowboy boots. He sliced our spiced ham, bologna and Longhorn cheese with his big, razor-sharp butcher knife, a knife that looked more like a small saber.

After slicing the meat and cheese, he then wrapped them in white butcher paper and tied them neatly with white string. He put them in a paper bag along with two loaves of bread and two candy bars.

Tom and I always had him cut extra slices of meat and cheese, so we could have a sandwich, our Pepsis, and our candy bars on the way home.

After crossing a bridge, we sat down on the side of the road on the levee facing the river and made our sandwiches while watching the farm tractors, pickups and cars cross the old, steel bridge that went over the Rio Grande.

We took our time going home, stopping every once in a while to throw rocks in the river or try to figure what it was that was floating down stream, or what had washed up on one of the sandbars that showed up when the river was low.

We walked down the levee to where the path went off the levee toward home. By then we could see the tin roof of the old adobe house and the outhouse, which sat at the end of the big cotton field that spread all the way up to the Richardson's place.

If Lila or Dave happened to be home, like when school got out early for them, which was usually on Fridays, they watched for us to come home and would let the dogs out to greet us as we lumbered our way down the path from the river. We always liked it when the dogs came out to meet us. The dogs running alongside of us kept the old rooster from charging and flogging us with his sharp spurs as we walked up to the house.

Usually after supper, Lila or one of the twins would help us with our homework. All three were good at helping us, especially with our arithmetic. Arithmetic was my worst subject.

Lila was good at art. She had a real talent for drawing Disney characters, especially Bugs Bunny in all his different poses and expressions, along with Elmer Fudd and Wiley Coyote. It was a shame that she never took it up in high school. She was also a majorette and looked very pretty in her boots and skirt, twirling her baton. And she was a fast learner on the piano that mother had managed to buy dirt-cheap from one of the cotton farmers. She took a few lessons, but had to drop out for lack of money.

Neither Dave nor Lila graduated from high school. Dave got a job working in a sawmill in the Black Range Mountains in New Mexico for the Inman Brothers. Phil Flint, a long-time resident of Canutillo who was an old friend of Laura's in-laws, the McCoy family, was also well acquainted with the Inman brothers who owned and ran the mill. He had hunted bear with them, and through his influence, Dave was able to get a job working in the mill, a trade he would practice until his death in 1995.

Phil Flint's wife was Post Mistress in the Canutillo Post Office, and their son Donnie who was my age, also went to Lone Star School. Phil was quite well known in and around Canutillo. He had a well-drilling rig and had drilled wells for many of the cotton farmers there. Phil was also an experienced hunter, and was very good at dressing out wild game. He led hunting parties into Arizona and New Mexico, and hunting expeditions into Mexico for eighteen years. His white stucco, Spanish-style home in Canutillo was like a museum, with display cases of Indian pottery, artifacts and arrowheads. The walls were covered with mounted bear, tiger, jaguar, and ocelot heads with pelts covering the floor from the various animals he had bagged in all these places.

We first met Phil when he stopped by the house with Laura's husband Johnny to drop off a bear roast from his latest hunt.

Tom and I were surprised when the dark green 1938 Chevrolet pickup pulled up in the backyard. We didn't recognize the driver; but when our brother-in-law Johnny stepped out of the pickup on the passenger side, we knew the man driving must be a friend of the family. As we started toward the pickup, we noticed that the pickup was rocking back and forth. We saw something moving in a big cage that took up the bed of the truck. By this time, Phil was out of the truck, checking out his other passenger—a small bear trying to get out of its cage. We couldn't believe our eyes, a wild bear in a cage in our backyard of all places.

Johnny introduced us to Phil Flint. And Phil warned us to stay away from the cage as the two walked toward the house. I had no problem with that.

Trip to Tonto Basin

While Laura and Johnny were spending the second week of their honeymoon in Arizona in the Tonto Basin north of Globe, Dave was home from the sawmill that had been shut down when Doug Inman, one of the brothers who owned and ran the mill was jailed.

It was a known fact that Doug had mental problems and it came as no surprise to Phil when he got word of the shooting, as he had known the brothers for years. When several sheriffs and Doug's brother talked him into handing over his 30-30 and he was taken into custody. He said that his reason for the killing was because the cook had been putting poison in his food. When the cook came out of the cook shack, Doug shot him just below his hatband, blowing out his brains, which spread across the ground and across the cook shack door.

Dave had no more than got home when Tom, Dave and I were invited to take a trip, a little vacation to the Tonto Basin with Corky, who was Johnny's mother, and Johnny's six-year-old brother Raymond.

When school was out, Mother thought it would be a good break for us to get away from the scorching heat in Canutillo and spend a week in the cool mountains above Globe. She talked it over with Corky. We would be going up where Johnny and Laura were spending the last two weeks of their honeymoon camping out. The last week of their honeymoon Johnny would get to go on a "Bear Hunt". We were traveling to the mountains in Corkys' 1934 International Carryall, the one she used on the rough dirt roads on the family cotton farm. The Carryall was used to run to town to pick up tractor parts and haul pickers with their cotton sacks and water cans to and from the fields on the big farm.

Our vacation ride would take us from Canutillo to Las Cruses and from there to Globe on Highway 70.

The Lee Brothers were going to take Johnny bear hunting to make sure he got his bear, They had hunted with Phil many times before in the Tonto Basin and were going to there with their hounds. Having hunted bear in the basin many times before they knew just where to find them. They were going to take Johnny on a hunt and let him shoot a bear. The week we would spend there with Corky would be the last week of Johnny and Laura's honeymoon. The Lee Brothers were going to use their hounds to tree the bear and let John shoot it.

After camping out a week we loaded all our camping gear and bedding in the old International we pulled away from camp with Raymond, Tom and I riding in the back setting on the mattress Corky had slept on with Raymond in the tent, the one we had sat on coming up. The trip home ended up a near disaster. We were lucky that we all lived to tell about it. After coming down the long winding road from the Basin we headed east on Highway 70 out of Globe going toward Safford and Las Cruces; from there it would be smooth sailing on home.

Corky was tired from the long, nerve-wracking drive down the winding road from Tonto Basin with four kids chattering in the car the whole trip. By the time we reached the little town of Duncan on the border going into New Mexico; she had to have a break. The little town, which we called "a wide spot in the road," is over a hundred miles from Globe.

We thought she would stop at the little park in town and rest in the shade for an hour or so, but she chose instead to pull off on the side of the road for us to stretch our legs after we crossed the border. Before we got back in the car, she asked Dave if he would like to take the wheel, if he would like to drive. She said she knew the road pretty well and it was pretty straight, but he would be driving over some hills first and there were places where there were some tight curves before we would come to some of the long, straight stretches. She said she would give Dave fair warning ahead of time if we were coming up on a sharp curve or a long hill. She figured it was safe to let him drive; there was hardly any traffic going either way. It didn't take much coaxing to get Dave to take the wheel. He had been driving but needed the experience.

After a short rest, we all piled back in the car with Tom, Raymond and me sitting in the back on a mattress surrounded by all the bedding and camping gear on the floorboard behind us.

Corky forgot to mention a very important fact to Dave about the condition of the steering system. The Carryall needed some work done on the front end. She hadn't told him that it would shimmy at certain speeds when it hit a big bump, and might be hard to keep on the road if both front wheels started wobbling.

He had no idea what was about to happen. We were doing about fifty when we hit the first big bump. The car started shaking and veering from one side of the road to the other with Dave and Corky both grasping the steering wheel, trying to keep the car on the road as the Carryall went out of control.

The next thing I knew we were all cartwheeling out of the car onto the shoulder of the road. We all ended up bleeding and bruised with skinned arms, legs, and elbows, with every bone in our bodies aching from being flung out onto the gravel on the side of the road— everybody except Raymond; he didn't have a scratch.

By some miracle he was screaming and crying, sitting on the mattress in the middle of the highway with camping gear, pieces of glass and debris from the wreck strewn all around him and for more than a hundred feet down the highway where the car had come to rest on its top.

Tom, Dave, Corky and I all ended up bruised and bleeding from the raw abrasions we got sliding down the pavement. Tom wasn't only bleeding from his skinned up arms, legs and elbows, but the back of his head was also bleeding from a deep gash he had gotten from some of the flying debris. We were real lucky that there was no oncoming traffic at the time the old car rolled. The wreck could have been much worse if we had collided with a car coming up the hill.

We stood on the side of the road, stunned, as Corky and Dave quickly dragged the mattress to the side of the road with little Raymond screaming at the top of his lungs.

We had just cleaned the debris off the road and were all sitting or lying on the mattress at the side of the road when a car headed for Lordsburg came by. The car came to a screeching stop and the driver jumped out and ran over to us to see if he could help, to see if any of us were critically hurt. After finding out that none of us were critical, he pulled out and headed for Lordsburg to send out an ambulance. It seemed like he was only gone twenty minutes when we heard the screaming sirens and saw the flashing lights of the ambulance and police car as they came up the hill.

We were really lucky no one was killed. In a short time, we were in the Lordsburg Hospital Emergency Room. As soon as we got there, Corky called home to let them know about the accident and to have some-one come and pick us up. She was treated for two cracked ribs and the gash in Tom's head was cleaned and bandaged; he didn't have to have stitches. Our abrasions were cleaned and bandaged and our bruises checked out, and late that night we were released.

When we finally got back home, we found out that Johnny's bear hunt had been successful. Laura proudly showed the pictures of the nice big bear Johnny had shot on their honeymoon camping trip in the Tonto Basin.

♪

There were a lot of things us kids could have gotten into if we had just had the money, like arts and crafts or music. There were several times that I wanted to take the bus trips with my class when they went to El Paso on Fridays to attend carnivals and different school functions. Even with everybody working, there never seemed to be any extra money for that kind of fun.

I did get to make a trip to El Paso one cold evening riding behind Doyle Lawrence when he was trying out his new Harley Davidson motorcycle. It was my first and only trip there on the back of a motorcycle. I thought of the trip for weeks after, dreaming of how nice it would to someday have a nice big motorcycle of my own.

Tom and I spent our weekends playing in the back yard, or playing in the homemade tent we had made from a torn cotton bale net that had fallen from a cotton trailer. We had stretched the net between some willows that grew between the levee and cotton field.

I enjoyed riding my bicycle too, a delivery bike that the twins had brought home for me while they were working at Western Union in El Paso. It was a good strong bike but needed some work on the brakes. It had different-sized wheels with the smaller of the two on the front; above it was fastened a big delivery basket.

My first job was at the Texaco station in Canutillo. Tom and I usually stopped by Brisco's Texaco or Lawrence's Grocery Store for a cold soda on our way home from school if we had a dime. Brisco's was the closest to home. One afternoon we had just pulled our Pepsi's out of the big red Coke box, popped the caps off with the opener on the side of the machine, and were enjoying them standing in the shade of the canopy near the side of the station when one of the owners, Buck Brisco, wearing his Texaco cap with the big star in the center, walked out of the Lube Room, and walked over to me to offer me a job.

He said that he needed someone to clean up around the grease rack, clean the toilets, sweep around the pumps and counter and empty the trash cans into the dumpster around back. He would pay me two dollars and fifty cents a week. I was really happy. Now Tom and I could have a soda every day and I'd have money left. I could hardly wait to get home to tell Mother. Now I'd have money enough for us to go to the movie in Anthony. I'd even have money to buy us a comic book!

There was no theater in Canutillo so we went to the closest one six miles north in Anthony, New Mexico, called the New-Tex Theater, named for being on the border of New Mexico and Texas. We rode the Greyhound Bus on Sunday afternoons between one and two when they had double features and sometimes two cartoons. I had plenty of money for round trip tickets, show fare and popcorn for both of us, and with the extra quarter we each got from Mother we were able to get us a comic book at the drugstore where we waited for the bus for our return trip home. Back in those days it didn't cost twenty dollars for Pepsi's, popcorn and a movie.

One afternoon on our way home from school, Tom and I stopped at the Lone Star Café. As we stepped in the back door into the kitchen we noticed Mr. Baker sitting near the pantry with his pant leg up bandaging a sore on his shin, a sore that never seemed to heal. Mother said the reason for that was that he more than likely was a diabetic and diabetics had to be careful about getting cuts or bruises; if the diabetes were bad enough, their wounds would never heal causing them to lose a foot or leg.

Jerry Baker Gets in Fight

We had just sat down on a couple of empty beer cases enjoying our sodas that Mother brought us when we heard the two Baker boys coming up to the back of the restaurant shouting obscenities at several Mexican boys, one in particular that was ready to fight the oldest boy Jerry.

The two Bakers had just opened the door to come in when H.E. standing by this time holding a large fork he had taken from one of the empty roaster pans sitting in the sink. He heard the Mexican kid calling Jerry out to fight. Just as Jerry started to close the door Mr. Baker shook the big fork at him, moving his lips a full minute before the words came out, saying to him you get back out there and fight that Messkun or you'll get this meat fork up your ass."

Jerry took his glasses off, went back outside and the fight was on. I sat there with Tom and drank my soda I knew if I went out I'd probably be right in the mid-dle of it. After about ten minutes the fighting had stopped. The Baker boys came back in and Jerry start-ed telling his dad what it was all about as he checked out the skinned places on his face knuckles and el-bows. The four of us sat there drinking our Pepsi's talking about how those Messkuns liked to start fights with the Baker boys talking in their Texas slang, talk-ing about "chunk'n" rocks at the Messkuns and about "tho'n stuff in the river and chunk'n rocks at it."

We finished our drinks, said goodbye to Mother and walked home. We certainly found out that Mr. Baker certainly wasn't going to have one of his sons back down from a fight. He was really funny to watch when he talked, his lips seemed to move forever before the words came out.

One Saturday, Jerry and Jackie invited me to pick cotton with them since they had extra cotton sack. It didn't take long to find out that I wasn't cut out for that kind of work.

I had no gloves and had a terrible time keeping up with the Baker Boys. By the time the day ended I was more than glad. With an aching back and both hands and fingers bleeding from the thorny cotton bowls, I ended up making less than four dollars. That was the one and only time I ever picked cotton.

It seemed like we never had a dull moment when Bill was around. If he wasn't telling some straight faced funny story or ridiculous lie or playing his guitar while purposely whistling off key, he was teasing the dog or one of us kids. He like to teas Tom when he was a little guy with a limited vocabulary. He teased him saying to Dave and I that he caught Tom "hesitat'n' behind a tree" or teasing him, saying things like "Oh no, you've got garments on your back, your back is just covered with garments."

Thinking he was covered with some terrible parasite Tom would run in the house to Mother, yelling at her to look on his back and see if he had garments; or embarrassingly telling her that he was not "hesitat'n' behind a tree," saying Bill was lying, he was just hiding.

If he wasn't doing silly things like that, he would be concocting some kind of explosive to test out on a can buried in the ground or some other object that was unthinkable for a bomb test, like something we used every day, like the old adobe outhouse in the back yard with its lop-sided door and crumbling bricks. He always had something going, always found a place for humor even in some of the toughest situations, like the one I'm about to tell.

Unexpected Visitors

One hot afternoon a couple of un-expected visitors showed up in the front driveway. When Mother answered the door she was greeted by a slightly built man with thinning hair wearing slacks and a sport shirt. He said that he was looking for Bill Creekmore's house. He said he had stopped at Lawrence's Grocery store and had followed Mr. Lawrence's directions to the adobe house in the bend of the road on the other side of the river.

He introduced himself as Dad's half-brother Leonard Creekmore (Leonard Shelly). He said he was traveling with another of Dad's ken, a second cousin by the name of Herman Land.

Tom and I looked out through the screen and saw the shiny new black 48 Chevy four-door sedans with

Tennessee licenses plates that had just pulled in and parked in front yard a few feet from the front door.

In the front seat sat a heavy bald headed man that was apparently sleeping against the half rolled down window on the passenger's side. Leonard said he and Herman had set out about five days ago from Jellico, Tennessee.

He said they were breaking in Herman's new Chevy heading for the west coast when they decided to stop by and see Bill on the way.

Mother told them that Dad had a job overseas on the Island of Guam and had been gone for several months. She had heard Dad mention their names faintly remembering Leonard; but she in all her years living in Kentucky had never met Herman.

She said her oldest son Bill would be home later on that evening returning from his bartending job in Las

Leonard said that Herman had some health problems and was suffering from neck and back pain and he sometimes turned to alcohol to relieve his pain. He said Herman's neck and back were a mess and the pain he suffered was from injuries he acquired playing football. He had played for Notre Dame in the mid and late thirties and had taken a coaching job before the war. Mother invited them in. Leonard had to help Herman out of the car and into the house.

It was quite a sight, Herman Land was huge, and must have weighed close to three hundred pounds. As they struggled helping Herman get through the door, we noticed what mess his clothes were in. What had once been a clean well starched and pressed white short sleeved shirt and once nicely pressed grey slacks, were now both wrinkled, rumpled, and stained with the crotch and right pant leg soaking wet. Apparently Herman had passed out and wet on himself.

Herman reeked of urine and alcohol. It was quite obvious that he was drunk. Mother and Leonard noticed that he was struggling to get his breath and complaining of chest pain he said he had indigestion. They thought he might be having a heart attack so they had him stretch out on Mothers bed where he lay until Bill came home to help Leonard get him to a doctor.

By the time Bill came home Herman's chest pain had gone away after Mother and Leonard had him drink some baking soda and water. Bill and Leonard discussed his condition and decided he needed a good laxative, something to clean out his bowels.

There was no hospital or doctors that we knew of in Canutillo and they weren't going to put him in the hospital in El Paso. Bill called Mother aside and told her their plan, he said he would take care of the situation and they would be back on the road the next morning.

He would go over to the little store, pick up something to break up that dam in Herman's intestines. He was gone about a half an hour and came back with the three common laxatives of that time, Ex Lax, Sal Hepatica, and Epsom Salts which is one of the best ones for fast results.

If none of these three worked, the only option left to un-clog that gut was for him to give himself an enema. This of course would have to be done in the outhouse since we had no inside toilet or a chamber pot, or anything of that nature for Herman's big butt. They hoped they would see positive results within the hour.

After drinking down a double dose of Epsom salts in warm water he still wasn't relieved, and by ten o'clock that evening after taking both doses of Salts and Sal Hepatica Herman still wasn't relieved so they stayed the night.

It was about seven the next morning when he woke everybody grumbling and stumbling around complaining to Leonard about his stomachache. Bill told Tom and I to take care of the rooster so he wouldn't attack Herman as he made his way from the back door to the outhouse while Bill prepared the enema of hot Epsom salts.

The old adobe outhouse was a good forty feet from the back door. It sat facing the cotton fields and the levee, with the door facing the river.

The door to the small one holer was sprung and stood half open, when in use the jury rigged latch on the inside was the only thing that held it closed, the broken spring hung loose near the top hinge. This was because several months earlier Bill had tested one of his explosive devices between some of the already loose adobe bricks. In doing so the explosion sprung the door and broke the already worn out spring that pulled the wooden door shut.

Since Herman was so big and wobbly on his feet it would be a tight fit to try to manipulate the enema inside the small toilet, there just wasn't room.

It was decided that Leonard would have to give him the enema while Bill stood at the side at the ready, holding and pressing the rubber bottle full of hot Ep-som salts while Leonard held Herman's shoulder keeping him from falling on his face as he shoved the enema home.

Just as the bottle had emptied and Herman turned, squeezing himself in, positioning himself over the hole, sounds of rapid Spanish being spoken between bursts of laughter wailed from the cotton field, as the Mexican cotton choppers, the unexpected on lookers, got a grand stand view of the hilarious scene.

Here right before their eyes was a huge, bald headed half naked man with his pants down around his ankles, grunting and holding his butt cheeks apart while being assaulted with an enema while two men, one bracing him to keep him from falling over, the other with his arms stretched high pressing tight against the bottle of hot Epsom Salts at the side of the ramshackled old outhouse. What a sight, what a comedy.

By this time Bill and Leonard, realizing the hilarity of the whole scene, were in stitches themselves. Herman's bowels exploded, giving him his much-needed relief. Trembling and pale as a ghost, Herman pulled up his pants and staggered back to the house as Tom and I kept the old rooster occupied while Bill helped Leonard load their suitcases in the car. We all said our goodbyes as they drove away, headed west. Herman Cecil Land, born August 19, 1911, Whitley County, passed away May 19, 1957, a young man of forty-six.

♪

Not far out of town, right off the highway going to El Paso, there was a little service station and roadside zoo run by Mr. and Mrs. Vail. Their little family zoo had been there a few years and they had the typical desert animals and birds. The little zoo had a pair of roadrunners, a hawk, an armadillo, a good-sized terrapin, and a large bobcat. The main attraction was a small monkey that kept everyone's attention. The little monkey had a habit of picking up a cigarette and puffing on it if someone flipped one into the cage that he shared with the terrapin. Flipping cigarettes into the cage was not allowed, of course, according to the posted warning signs.

It was quite funny watching him take a few puffs off the cigarette and then try to share it with the terrapin that pulled his head in leaving the monkey poking the cigarette at the shell.

One day Bill stopped in to get gas on his way to visit a visit a friend that lived not far from the zoo. As he came out of the station, he noticed Mr. Vail carrying the bobcat that had died. Bill asked if he was going to skin it and save the hide. Mr. Vail asked Bill if he was interested in the hide, and told Bill he would gladly give it to him if he wanted it. So Bill brought it home.

That afternoon as Bill was skinning out the cat in the back yard, one of the Wells boys, the younger one, happened to be on his way to town riding his bicycle. When any of the Wells family walked or rode their bicycles they usually went down the trail to the levee, which was a shortcut to Canutillo.

The gravel road on the levee had very little traffic and they walked or rode down the trail passed the outhouse toward the river. Since there was no fence that closed in the back yard where the circle driveway come off the pavement and circled the house made it easy to walk or ride into the back yard if you didn't have to fight off the old mean rooster.

When the Wells boy saw that Bill was skinning out the big bobcat, he got off his bike and came over for a closer look. Tom and I noticed that he had a dip of snuff in his lower jaw when he said hello. He walked casually pushing his bike toward the small table where Bill was skinning out the cat.

He tried to act more like a grownup than the kid he was. He turned his head, spit and struck up a conversation with Bill as he stood there leanin' on his bike. "Where did ye get that big cat?" he asked Bill, spurtin' out snuff in between his questions.

Bill, the storyteller that he was, said he got the big cat when he was camped out in the rock quarry up on the mountain behind the cemetery. He said, "When he got too close to my sleepin' bag I had to shoot 'im."

Bill asked how he liked the rooster, how the family was doing, and a how his grandpa, who had just re-cently moved into the little white stucco house in the cotton field with them, was doing.

The Wells boy cleared his throat and spit, and looked real serious as he wiped the snuff spit off his chin and told us about the pain his Grandpa was going through after having several of his front teeth pulled.

We busted out laughing when he blurted in a very serious, "He cain't eat nothin'. Now his mouth looks like a horse's ass with a sore in it," and spit again.

He said the rooster was really good, that they "had 'im for dinner a couple of weeks ago." Having said that, he put another dip of snuff in his mouth, said goodbye, and left the yard peddlin' down the path toward the levee.

Losing Tippy

Dad came back from Guam the end of March in 1948. Earlier that month, March the sixth, our much-loved little dog Tippy was killed crossing the bridge into Canutillo. One of us had left the door ajar and he had gotten out, following us to town; he got as far as the bridge.

We had searched all over for him, checking with the Wells family and our landlord Mr. Richardson to see if he had followed one of their dogs home. We walked down the levee calling his name. We stayed awake half the night worrying about what could have happened to him.

As Tom and I crossed the bridge the next morning, we got about half way across when we saw him lying on a sandbar near the river. He had been hit while crossing the bridge. With tears running down our cheeks, Tom and I took him home. When everyone got home that night, we were one sad family. It was like losing one of the family.

Shortly after we lost Tippy, Dad came home from Guam. He had a month off, a month's vacation. He had planned on the family getting the necessary shots and papers to make the return trip with him. He had plans for us to move to Guam when he returned to work.

Mother said she didn't want to move that far away from the family. She had been writing to Aunt Minnie who was now living in Burns, Oregon where they had been for the past two years. Aunt Minnie said that there was plenty of work in and around Burns, and she said that Uncle Grant had been working steady at the Chrysler Plymouth dealership doing body and fender work. He could be influential in getting Dad hired; there was a mechanics job that would be available in a couple of weeks.

Uncle Grant was real good at his job. He could make nice cars out of some of the worst wrecks. He was an excellent painter and was real good at matching paint by blending colors to match the original paint. Back then there wasn't the computerized technology that they have now for mixing and matching paint. Matching paint in those days wasn't that easy, and a lot of painters didn't have that talent.

Aunt Minnie said that Dave and young Grant both had jobs working at one of the several sawmills. She thought Bill and Dave could probably find work too.

After talking it over they decided that it would work out the best for us if we moved to Burns. Dad got a letter off to Guam telling them of his decision to stay and requested the back pay he had coming.

He bought another old Oldsmobile, an ugly, faded green two-door sedan with ratty upholstery, no back bumper and a transmission that popped out of second gear when down-shifting or when going up a hill. The way the light green primer showed through the topcoat of paint in painted stripes, made the car look like a huge watermelon.

The engine smoked but didn't have any bearing noise or knocks. Dad took a four-by-four fence post, cut it the width of the car, and bolted it on as a back bumper. The front bumper had already been replaced with a four-inch water pipe. We removed the back seat to make room for boxes of tools, dishes, and pots and pans, and on top of those we put a mattress and the box of utensils we kept out for cooking along the way. When nightfall came, we camped along the way as we had done before.

Abbey, Tom's dog, the only dog left after losing Tippy, was limping from being hit by a Greyhound bus in Canutillo, right across the street from the Lone Star Café. He too had the ill fate of following us to town.

Mother was coming out of the café kitchen with a tray of food when she saw several people standing, looking out the big front window at a yelping dog, writhing in pain, dragging his right hind leg. He dragged himself toward the restaurant after being struck by the passing bus.

Mother recognized him right off. She knew the dog was poor Abbey. With the help of one of the customers, she took a large cardboard box and carried him to the back of the restaurant. She thought he was suffering internal injuries also. She rigged up an ice pack and placed it around the damaged leg and got him to drink a little water.

When she brought him home that night, none of us thought he would live. Tom was in tears. None of us expected him to be alive the next morning, but he fooled us all. He was able to eat, and with Tom's help, a few days later, he was able to stand. He was twisted and limped when he walked, and he could not put any weight on his right leg. We later found that his hip, not his leg, was broken. After about three weeks, he was able to hobble around on three legs, but was unable to run.

Being short on money as usual, we never did take him to a vet. It would have broken Tom's heart if we had given him to one of his tender-hearted school mates, or worse yet, have Bill put him out of his misery by shooting him behind the ear, a brain shot, like so many people did in those days to relieve their animals in pain.

Trip to Burns

On our trip from Canutillo to Burns we stopped again at Blythe to rest and clean up in the river. I am easily embarrassed and make sure never to expose myself to anyone when I'm in the shower or getting dressed.

I had just stepped out of the shallow water and was walking toward my clothes on the bank when all of a sudden Bill and Dave popped out from behind some willows and saw me.

Bill looked at Dave, smiling, and said, "Man, look at all that hair growin' around Jim's moose." I could feel my face burnin' red as I hurried to get dressed.

The rest of the trip was long and uneventful, except for the thrill Bill and I had when a tire blew out going down a fairly steep grade in Nevada.

The tires on the old Olds' were not the best. The tread on two of them, or what was left of it, had worn spots that had the cord showing through. They were more like what I would call "tire-less tubes," and the spare wasn't much better. Most of the tread was gone on the rear ones and the right front had a slight bulge in the outside sidewall. We had just started down the long grade doing about fifty miles an hour when that bulge blew out the sidewall. Bill was having a tough time keeping us on the road after it blew. He tried downshifting into second gear several times because the brakes were not slowing us down all that much and it kept popping out of second gear. It might have helped if the car had power steering, but no such luck. He had to keep it from pulling us over the steep bank to the right.

I made the mistake of pulling the emergency brake. The emergency brake handle was located right by the transmission on my side and made it easy for me. I reached over and jerked it back. Instead of braking evenly on both rear wheels, just one wheel, the left rear, locked up burning rubber for a hundred feet as the old car was trying to switch ends. Bill was now having an even harder time keeping us on the road. He noticed what I had done and quickly reached over and released the emergency brake. By then we had slowed down enough for him to gain control and we came to a stop on the shoulder.

He was mad at me at first, but his anger turned to laughter as we scotched the wheels and changed the tire. We joked about how funny it would have looked to the people in the cars coming up the hill. If we had rolled, they would wonder where this giant watermelon rolling down the hill had come from!

Before we left Canutillo, Dad had mapped out the highways with the least mountains to cross, or so he thought anyway. We traveled as far north as Ontario, Oregon, which sat on the other side of the Snake River, and from there we took Highway 20 on into Burns some 130 miles west. We had one more steep grade to go over between Ontario and Burns–Stinkin' Water Pass–and from there it looked like a pretty easy ride on into Burns. We had to stop several times to fill the radiator. Between that and having to run in first gear because the second gear was useless, it took us twice as long to get over the mountain.

After the long slow climb we finally reached the summit and started down the other side. Tom and I both let out a sigh of relief as the temperature gauge began to drop and the groaning transmission quieted down. There wasn't much to look at as we started our descent, just gates a little ways off the road with long gravel driveways going to the various ranches.

Water seemed to be running everywhere from the patches of snow and slush higher up that were melt-ing. There was water gushing out between the rocks, and in several places we had to swerve to miss large rocks and debris in the road where it had tumbled down the slopes. The state highway maintenance crew had pushed some huge boulders to the sides of the road.

We were talking about how lucky we were not to have been there when the real big stuff, the boulders, had rolled down. In several spots we had to drive in the left lane to miss the mudslides and big rocks. Apparently the landslides had happened before daybreak that morning.

Dad and Mother, following in a second car, caught up with us when we were flagged to a stop farther down the road. It was stop and go for several miles as they cleared the road. We finally reached Buchanan, a little service station and gift shop, which had a big jar full of some of the biggest rattlesnake rattles we had ever seen. We filled the gas tanks and radiators and headed out for Burns, now a little over twenty miles away.

4 AUNT MINNIE'S

Harney County in the spring was quite a sight with all the wildlife. Water was everywhere and the fields were flooded and loaded with all kinds of migrating birds. There were huge flocks of snow geese, Mallard ducks, and cranes. And in the drier fields were herds of antelope and deer feeding on the tender shoots of grass between the patches of snow. We hadn't seen anything like this since we left Colorado.

The thousands of acres of wild hay that filled the fields, along with the alfalfa the ranchers planted after the snow melted made it an excellent place for raising cattle and sheep. By midsummer, the big rigs loaded with cattle, sheep, and bales of hay were a common sight in and around Burns. When we moved there in the spring of 1948, there were also lots of logging trucks pounding the pavement coming from the Ochoco or Malheur National Forests. The logging industry was booming, and jobs were plentiful if you were willing to work.

There were several small sawmills near Burns, and one of the biggest in the country was the Edward Hines Lumber Company, just a mile or so west of Burns in the little town of Hines. We couldn't wait to get there.

After crossing the railroad tracks near John Day Junction, the houses and ranches grew closer together, but we still didn't see the town until we broke over the last hill. There it was, downtown Burns. We pulled into the first filling station we saw to get directions to the Carroll's house on Riverside Drive. Dad introduced himself to Red Wolverton, the manager of the filling station, and after getting directions; Bill called Aunt Minnie and told her we were on our way.

We followed the gravel road past various homes, both large and small, some two stories, some of them built in the early thirties and before. Many houses had nice front lawns and shade trees with some un-melted snow in the shady spots near the stacks of firewood in the backyards. Just about everyone burned wood for heat at the time, and they also used wood for their cook stoves.

A plank sidewalk ran in front of the homes on one side of the street, which was lined with tall poplar trees. The sidewalk went more than halfway down to a bend of the road where it ran along the tangled, wil-low-covered banks of the Silvies River. A little one-lane bridge crossed the river in at the bend, and dead-ended in the Carroll's driveway.

We eased over the bridge with its heavy planks and two-by-four handrails, the swollen river not three feet below. As we came off the bridge, we could see the little ranch with its flooded fields near the barn at the back of the property. Water was everywhere there was a low spot on the property, even some between the stepping-stones leading up to the front porch. What a difference from the place we had been living. The green fields were not like the irrigated green fields of cotton that surrounded our house in Canutillo, and certainly not like the dried-up adobe yards full of prickly pear cactus and ocotillo.

Aunt Minnie and the kids, Donnie, George, Dave and Nancy, and their dog Big Paws greeted us as we came to a stop in the front yard. Uncle Grant and young Grant weren't home from work yet, but would be home soon. The Carrolls were really happy to see us.

Aunt Minnie and Mother enjoyed their dips of snuff while sitting on the small front porch, catching up on the latest family affairs while we kids took turns throwing a can lid for Big Paws. He dashed out through the ankle-deep water to retrieve it, and after being scolded for getting the dog so wet, we went into the house to check out their latest comic books and Big Little book collections.

When Uncle Grant and Grant Jr. came home from work, we sat down at the dinner table and enjoyed Aunt Minnie's huge hamburgers with plenty of mashed potatoes, gravy, biscuits and beans. We topped it off with her delicious homemade wild blackberry pies from the berries she had canned the previous summer.

We stayed with them for two weeks and then rented a small, three-bedroom house in town, not far from the main street of downtown Burns. It was an old house with high ceilings and overhead lights with long string-pull switches. The old house had two small porches, the front one was made into my bedroom later on. We lived there while I was going to grade school and through my first two years of high school.

The old town of Burns, including the little mill town of Hines, had a population of about 5000 residents. In Burns, the homes were mostly from the twenties and thirties. The houses in Hines were more like tract houses, built and sold or rented to the people working for the Edward Hines Lumber Company. There were plenty of jobs to be had if you didn't mind hard work.

There were several small mills like the Wolverine and the Pinecone, to mention a couple. There were jobs to be had working on the various cattle and hay ranches in Harney County. Some paid up to eight dollars a day and included room and board. They all needed tractor drivers during haying season to drive the bright green John Deere "Pop'n Johnnies," one of the more popular tractors used to pull the hay rakes and mowing machines, and to disc up the huge fields for planting alfalfa.

There were always jobs for buckaroos and all-around ranch hands, those who could do anything from milking cows to breaking wild horses. There were jobs too for Gandy dancers, the men who worked maintaining the railroad tracks, laying ties and driving spikes, which was a never-ending job, especially with all the freight trains coming and going from the mill towns of Seneca, Burns and Hines. The bars and eating places were filled with loggers, cattlemen, mill hands, and the local business owners.

Burns, like all small towns, had its characters and Bill, after tending bar at several of the bars, became well acquainted with some of them. He was always telling us funny stories about some of them who frequently came into the Palace, the bar and restaurant where he worked most of time.

Some characters had nicknames like "The Corpse." The Corpse was a woman in her late fifties, who looked like a cadaver with spots of her pale skin cov-ered with pancake makeup and eye shadow on so thick on her pallid, bony face that she looked like a character out of a horror movie. Tom and I got to see her one night when we stopped by the Palace to get show fare from Bill for a movie we were going to see at the Burns Theater down the street.

There was another popular character called "Stuttering Jess," an alcoholic about thirty-five years old. Bill said when Stuttering Jess got loaded after four or five mugs of beer, he had a real tough time talking without stuttering. When he tried to order more to drink or carry on any kind of coherent conversation, he would stutter so bad it was difficult to understand what he was trying to say. I got to see and know these characters after Bill got me a janitor job at the Palace.

There were two elderly brothers who often came in, Gilbert, the oldest, and his younger brother Carl. The two white-headed Swedes loved their beer. Gilbert, a pathetic World War I Veteran had one crossed eye and a flattened, broken nose, like someone had hit him in the face with a shovel. He was a real embarrassment to his younger brother Carl. His actions were especially embarrassing when he got loaded and after very little coaching from Bill would start singing "Sweet Adeline" at the top of his voice. His off-key voice could be heard over the rest of the noisy bar crowd, especially with Bill directing him by swinging his baton (a small toilet bowl plunger that he used for unplugging the tap drain behind the bar). When this happened, Carl usually moved to the other end of the bar, to get as far away from his brother as he could.

As Gilbert sang "Sweet Adeline" at the top of his lungs, he sometimes dropped the upper plate of his false teeth on the barroom floor, and Carl would be sting at the end of the bar staring at his beer glass trying to ignore the whole show.

Carl, who was quite eccentric, addressed himself as "Carl the Snow White Viking" and when Carl came in alone, Bill would make him furious by pretending to mistake him for his older brother Gilbert. "What would you like to drink, *Gilbert?*" Bill would ask him. Carl would get mad. "I'm not Gilbert. I'm Carl the Snow White Viking. That son-of-a-bitch is my brother."

When I stopped by after school to do my janitorial work, I got acquainted with Emery Lanfear and Francis Griffin. These two characters were just as amusing as Carl and Gilbert.

The Wagner brothers, Bill, Adolph, and Fred owned the bar, and one of them would be in the bar every day. They were goodhearted, and welcomed most everyone who came in for dinner and drinks. Emery Lanfear was a frequent customer at the Palace, but Frances Griffin didn't come to town very often. He didn't drink, and didn't stay in town very long when he did come. He usually came to town to grocery shop and to fill his feed sack with scraps from the garbage cans in the alleys behind the grocery stores, butcher shops, and restaurants. He said he was going after food and bones for his dogs.

He would come to the Palace to eat and enjoy looking through the magazines in a big rack in the sporting goods section near the front of the bar.

He would buy a quart of milk and a loaf of bread at Wenzel's grocery store, take them down the street to the Palace, find an empty booth, sit down, and draw a big spoon from his overcoat, drink down the water that was set before him, break up the bread, mixing it with the milk he poured into the empty water glass and sit there until he finished the quart of milk and half the bread. He would then put the empty quart milk bottle in his big overcoat pocket, go over near the front door by the magazine racks and would say to the customers coming in, "Kind sir, would you be so kind and give me some change?" After collecting a few dollars, or after Bill or one of the Wagner Brothers made him go outside, he would head back to Wenzel's to return the empty milk bottle and get back his deposit.

When the weather was bad and the restaurant wasn't crowded and there were plenty of empty booths, he was allowed to come in and enjoy his milk and bread in one of the booths. He was clean-shaven and wore the big overcoat both summer and winter; it served as his blanket when he sacked out under a bridge or wherever he found shelter for the night.

One day at school, one of my schoolmates was telling me about Frances and of his involvement in the yearly parade that took place in Burns a couple of years before we moved there.

Apparently Frances had his own small ranch and had raised some prize white-faced bulls, and he brought one downtown to take part in the parade. He had the fine-looking bull on a halter and followed behind it on the back of an old, skinny, sway-backed horse.

As he came down the street, crowds on both sides of the street began to cheer and clap their hands as he rode slowly by. He was doing fine until someone in the crowd was nudged a little too far out in the street and it spooked the bull. Frances tried to hold the bull by halting the horse like he did on the range, not realizing the horse was trying to dig his hooves into slick pavement. The crowd began to roar with laughter as horse and rider were dragged by the bull to the end of the street.

Frances lived quite a ways from Burns. The last time I talked him, he said he was living in a "Buckaroo Cook Wagon" at Wrights' Point. I could hardly keep from laughing out loud when he was telling me how he en-joyed his evenings sitting in his "Buckaroo Cook Wagon lobby," watching the birds and antelope when they came to the lake to drink. It was said that Francis was a very intelligent man and a college graduate.

Emery wasn't an alcoholic like many of the others that would be waiting at the door at opening time. He usually came in on weekends in the afternoon or early in the evening. Bill opened the bar at nine o'clock in the morning and started off the day by not charging customers for their first glass of beer. Serving their first beer free of charge and taking time to listen to their stories made repeat customers out of the biggest part of them.

Emery would lumber up to the bar wearing his blue chambray shirt, black Frisco jeans held up by suspenders and his Stetson pulled down low, almost covering his right eyebrow, his half-shaven face, covered with patches of whiskers and shaving cream and nicks and scrapes where he had gone a little too deep with the razor. With a very serious look on his face, with eyes almost squinting, he would say to Bill, "Curly, draw me a beer."

As Bill set his glass down, he would ask Emery how everything was going at his mine. This of course was exactly what Emery wanted to hear.

Emery would start out talking about how deep into the mine he had been, telling "Curly" about the different ores he would find as he dug deeper. He had his own terminology in describing his mining, like saying, "I was down about a foot and hit Buckskin Brown, and about eight inches farther I run into some pockets of quick." meaning quick silver. He had saws and tumblers and a polisher for processing some of the rocks from his mine and would bring samples with him to the Palace.

One afternoon on my way home from school I stopped by the Palace to see if the Wagner's had any extra work for me, I needed all the work I could get since I had started saving money to buy a car. I would check with Bill to see if the Wagner's or anyone who needed yard work done or if they needed wood stacked.

As I walked in I noticed Emery talking to Bill at the front end of the bar near the sporting goods counter.

As he talked I noticed that he was reaching down to his right front pocket of his Frisco Jeans trying to get something that was tucked inside his pants. We watched as he folded his jeans back and removed a big horse blanket pin that was pinning the large coin purse that had been out of sight.

He had been telling Bill about some slices of "Picture Rocks" he called them that were in the big snap purse. Bill and I watched as Emery un-snapped the purse and drew out a neatly cut and polished slice of rock. He handed it to Bill. As Bill held it up to the light, Emery said "You see those animals in that picture?" Bill said, "I see what looks like some kind of an animal standing in a stream."

It was all we could do to keep from busting out laughing when Emery said, "That's a raccoon wading up a crick with a white rat on his back." He put his picture rock back in the purse, tucked it back inside his jeans and secured it with the big horse blanket pin. Emery worked at the Hines Mill and worked his "claim" on weekends.

My first job in Burns was working at the bowling alley, which wasn't but a few blocks from the house. I found out about the job while playing ball in the little city park. I was told that the Bennetts, Cleve and Sally, the owners of the Burns Bowling Alley, were always in need of pinsetters. Cleve Bennett was a veteran fighter pilot in World War II.

Setting pins was kind of a dangerous job because some of the bowlers threw the ball so hard that when it hit the pins, they flew up and hit the setter above the pit where the pins normally landed. The bowling alleys at that time had no automatic machinery so the pins had to be gathered up by hand and placed in a triangular rack that was lowered to position in the alley. After working there a week, and getting hit several times on the shins by the flying bowling pins, I knew who threw the balls the hardest, and I was lucky the flying bowling pins didn't break any bones.

I worked there through the biggest part of our first summer in Burns. That summer I was really proud of my accomplishments. I was able to buy my own school clothes and shoes. I was especially proud of my first pair of field boots with the big buckled strap across the arch, and my Levis and a fancy belt. The belt was covered with shiny studs, and between the studs were colorful red, green, and yellow plastic jew-els. They were like those on the Roy Rogers cap pistol belts and holsters we used when we played cowboy and Indians in our younger years.

It was night and weekend work and I usually walked home about eleven-thirty. I always had money for show fare and comic books that I always shared with Tom. After school started that year, I only worked for a couple more weeks. By then I had managed to buy a nice, used Rollfast bicycle, which I rode on weekends with some of the guys in my school class.

There were a lot of good places to go bike riding. The highways in and out of Burns were pretty safe for bike riders. We had no special bike paths, but we were pretty safe as long as we kept near the shoulder of the road. When we peddled down the long stretches of highway, we were able to see the big cattle and hay trucks coming and get off the road before they ever got close.

I always managed to have some kind of job after school. And I took various jobs on weekends. I might be digging holes for septic tanks for Lee Frazier, the local plumber or stacking firewood or "mill ends" as they were called. Just about everyone in Burns and Hines and the other logging towns used wood burning stoves for both heating and cooking. With Mother kicking in five dollars, I eventually saved enough money to buy a used Powell motor scooter called a "Doodle Bug." It had tiny wheels and a centrifugal clutch-type transmission. I guess it got the name "Doodle Bug" from the rounded cover over the engine and transmission that also held a passenger seat for a little one. The cover gave it the Doodle Bug look.

It was a Doodle Bug all right; it was gutless. It would do all of thirty miles an hour on flat ground, but when I had to go up a fairly steep hill, I had to get off and push while still holding on to the throttle. I made payments to Mr. Dickenson, the father of a school classmate, Ralph Dickenson. I rebuilt the engine and installed a new clutch band, but even after it was re-conditioned, it was still a gutless wonder. My first motor scooter was a real disappointment. Needless to say, I didn't keep it very long. I made my one and only long trip with it to Five Mile Dam and back. The bumpy dirt road with big potholes beat out by cattle trucks and tractors was no place to ride the little Doodle Bug. The scooter was built for flat, paved streets and sidewalks. Shortly after the trip to Five Mile, I sold it to one of the kids in town.

My years in both grade school as well as at Burns High were never boring, especially during my junior and senior years. There were four classmates, Bill Tolliver, Frankie Osa, Bill Randt (Bill Bergert), and Butch Whitman that I clowned around with. We were always clowning around and getting into trouble. We spent a lot of time in study hall, and since I was the only one of the five not playing sports, I spent many days after school helping Mr. Taylor, the high school janitor. If we weren't interrupting class by teasing and pulling jokes on one of the girls or guys, we were teasing one of the teachers. Our math teacher Mr. Robe was one we teased the most. Mr. Robe, considered by some to be a mathematical genius, had a short temper when it came to students cutting up in class.

One time the four of us came to class, and because we weren't expecting a test, we had left our notebooks in our lockers. We asked Mr. Robe if we could go back to our lockers and get them. He flew mad, snatched the calendar off the wall, and made us use the backs of the sheets as our test papers. He became even more furious when he started writing out the test problems on the blackboard, and we started snickering.

He stopped his writing, turned around and told us to quiet down unless we wanted the "skinned nose and knuckle treatment."

Mr. Caruthers was another one we harassed. He taught English and Dramatics. Our Dramatics class was held on the third floor of the building where there was a stage. The room was also used as the music room with drums, cymbals, piano and a bass horn sitting on the stage.

After finishing our class on the second floor, Frankie, Bill Tolliver and Bill Bergert and I went up the stairs to the band room. We always got there before Mr. Caruthers; he had to come up two flights of stairs from the first floor. I was usually the first one on stage and I would start banging on the piano. I was soon joined by the other "musicians," Tolliver, Osa, and Bergert, as they started pounding on the drums, clanging the cymbals, and blowing the big bass horn.

When one of the students on lookout signaled to us that they could hear Mr. Caruther's feet pounding steps as he ran up the stairs trying to catch us in the act, we immediately stopped. By the time he flung the door open, he was as pale as a ghost and all out of breath. When he walked in, all of us would in our seats as if nothing had happened. It was all we could do to keep a straight face and not burst out laughing when he flung the door open and stepped inside the room.

♪

I started football practice in my sophomore year, but dropped out after my second week so that I could work after school. I wanted a car and the only way to get it was to work after school. I joined the boxing team, which really helped keep me in shape. The workouts were in the evening and it didn't interfere with the jobs I had after school. I continued working out, hoping to have enough fights to qualify me to get into Golden Gloves. I stayed in pretty good shape that way clear up until I got my draft notice and decided to enlist in the Navy.

One night during that time, I stopped in the little logging town of Seneca and got into a fight. I was on my way back from John Day where I had left my Cushman Eagle motor scooter with the Cushman dealer to be sold. I was traveling back to Burns with a friend when we decided to stop at the Grange Hall in Seneca where they were giving a party for some of the foot-ball players that lived there. As usual, I drank a little beer and danced with several of the girls.

I didn't know it at the time, but at one point I was dancing with a girl that had a very jealous boyfriend, a big burly football player. When the band stopped playing and I was walking toward the parking lot, all of a sudden the big jealous football player spun me around and punched me, knocking me down between two cars. I was helped up and walked to my friend's car with blood running from my nose, a split lip and skinned-up face. It didn't take much talking to keep me from going back it to try to finish the fight. I was pretty well beat. We got in the car, opened us a couple of bottles of beer and headed for Burns. It was late when we got home. Mother and Bill had no idea that I had been beat up until they saw my bloody, scratched up face the next morning.

Anytime I boxed, I had to make sure I kept my mouthpiece in place because I had a sharp front tooth that would split my upper lip even with the slightest punch. It had happened to me when I was in the ring with Stacy Gibson, one of my classmates. Stacy won the fight when the referee stopped it because of the blood pouring out my mouth from his second punch after my mouthpiece had fallen to the canvas.

♪

There were a couple of girls I had crushes on while I was going to Burns High. The first was Darlene Ries, a pretty, dark-haired girl that I met one Saturday afternoon when I went swimming at the Hines pool. The other was Darlene Tiller, a local girl from Burns.

The Hines swimming pool was a big rectangular tank made of redwood planks, which were sealed and coat-ed with numerous coats of silver paint and was filled from a hot, artesian well that bubbled up just beyond the big Hines lumber mill. I rode my bike out from Burns along with some of my classmates to go swimming in the pool on the hot weekends.

Darlene Ries and her family had moved to Burns from Bellingham, Washington. She was a grade ahead of me, and already dating a guy in her class. I didn't have a car, so I really didn't have a chance. I did manage to buy a car before I tried to date the second one, the local girl named Darlene Tiller. The pretty girl's parents ran a grocery store and meat market in Burns. I was quite let down when she backed out of our date at the last minute. We were supposed to go to a graduation dance, a dance being held for the graduating class ahead of mine. She had accepted my invitation a week before and I had gotten her a nice bouquet of flowers, but when I called to tell her when I would pick her up, she said she had changed her mind and wasn't going.

Her parents had decided not to let her go. They were quite religious and they knew I drank beer with classmates at some of the parties; they didn't want to take the chance of letting their daughter go and maybe end up drinking beer. Looking back, I really couldn't blame them.

I received my draft notice in August 1953. I thought my grades were high enough that I would be exempt. I was planning on enrolling in O.T.I. (Oregon Technical Institute), a trade school in Klamath Falls. I had planned that someday I would have my own body and fender shop. I had already had some experience in auto body work at local body shop with Claude Funk helping him prepare my '49 Ford coupe and several others for paint jobs.

I just knew my grades would be high enough to get me into O.T.I. I knew the kind of money my uncle Grant made doing body and fender work. He had moved to Riverside in southern California a couple years after we came to Burns and was working in a body shop down there. A couple of the faculty members from O.T.I. showed up at Burns High in early January to tell us what the trade school was all about and to talk about the possibilities of getting jobs when we graduated. We were told that most of the students who graduated from Oregon Tech were almost certain to be placed in jobs after completing the course. We were told of the nice facilities they had for the students and how we could cut costs by working at various jobs in and near the trade school.

When it came to selecting the students for the different trades, it all narrowed down to the grades on our graduating report cards. This eliminated me, and several others. I talked to the teachers about bringing my grades up, taking tests and getting an early diploma. They said there was no way since I had to sign up within the next week or wait until the following year. This was impossible since I already had my draft notice.

Being drafted into the Army was not for me. Richard Canning, one of my classmates, and I drove to the Navy recruiting station in Bend where I signed my enlistment. Richard decided against it and backed out. He said he would join later but never did enlist.

I think my last three years in Burns were the most fun. Thinking back, I should have taken my schoolwork more seriously. In May 1954, I left for boot camp.

I said goodbye to the family the night before I left as Mother stood with tears running down her cheeks. She hugged and kissed me as I climbed aboard the 11:45 Trailways bus bound for Portland where I was to take my physical. If I passed the physical, I would take the train from there to San Diego where I would go to basic training or "boot camp."

Basic Training

There were only a half dozen people on the bus when it pulled out of Burns, so I had pretty much my choice of seats, and after riding for a while behind the driver, I moved to the back of the bus and stretched out on the big back seat.

I tried to sleep but I had too much on my mind. I already missed my bed and being home with the family, and by the time we got to the first stop, which was Hampton, the halfway point between Burns and Bend, I was sitting back up front talking to the bus driver. We stopped in Hampton for a fifteen-minute coffee break and rest stop. By then I fully realized I was totally on my own.

All I had were the clothes on my back and my shaving kit. Several more people got on the bus and I moved. Taking my shaving kit from the overhead, I moved near the back of the bus again, this time taking a window seat on the passenger side. After settling down with no one to the right of me, I began to think of all I had left behind.

I had left my car at the Ford garage for the dealer to sell. I made an agreement with Mr. Weeks who owned the Ford dealership; he would sell my car and later on, I would buy another car from him with the money from my car as a down payment on a later model. Or, if chose to later on, I could have the money mailed to me.

I kept thinking about Mother and my little brother Tom and wondered how he would do without me being there for him in high school, how he would get along. It was the first time that I was leaving the family and going so far away from home, and I was going someplace other than to stay with family like my aunt and uncle, a place where I could come home from if I got homesick.

After blood tests and a thorough physical that lasted two days, I spent one more night in Portland. We were to finish up our paper work and board the train the next day. They didn't have a barracks for us, and we were issued rooms in one of the older, second-class hotels, not far from the place where we had our physicals. There were two bunks per room with no air conditioning. Anybody that's been in downtown Portland on a summer night knows how hot and muggy it can get.

The rooms were small. There was a restroom and a drinking fountain with a little paper cup dispenser at the end of the hall. There were no restrictions as to coming and going, but most of us had very little mon-ey to spend. All I had, including my change, was about eighteen dollars that I hoped would last me until I finished basic training.

The hotel wasn't far from a greasy spoon restaurant on the same side of the street. It had quite a few customers. Russell Thiess, my roommate, and I left the hot room for a Coke and something to eat. After buying a couple hamburgers and Cokes and talking to some of the other guys from the same floor, we got us each another big Coke and walked back up the dimly lit street to our room. By then we thought the room would have cooled down. It was still sultry hot so I went over to the window facing the street and I was able to open the big tall window to let in some air. I stood with my hands on the windowsill looking down at the people passing below.

Thiess came over to the window to see what I was looking at, and we both started dropping chunks of ice and water on the heads of the people as they passed below. What a surprise, we cracked up as they grabbed the top of their head, running their fingers through their hair, wondering where the sudden splash had come from, or wondering if a bird had messed on them. Thiess was a cut-up all through boot camp. I had no problem with boot camp discipline and I was in pretty good physical shape from my workouts when I was boxing. I didn't have any problems with calisthenics. I had been running five miles a day and doing lots of push-ups and sit-ups. I had planned on trying out for Golden Gloves in the fall.

The summer months weren't much fun working out and doing drills on the "grinder" in the blistering heat in San Diego. The blistering days and the hot sultry nights were pretty rough. We had daily marching drills using World War II 30.06 rifles. We drilled for hours practicing the rifle drill called "Sixteen Count Manual." The Drill was a series of moves using the rifles.

We had inspections and our uniforms had to be snow-white clean and folded a certain way, our socks rolled a certain way, and our mattress covers had to be stretched skin-tight over our mattresses. If one did not pass inspection, the whole company suffered. It meant more push-ups or other hard exercises.

Sometimes after evening chow the teasing and insulting remarks would cause fights to break out in the barracks. One such fight broke out one evening as the barracks bully slugged it out with a couple of the weaker kids. I stepped in when he started fighting with Casey, a tall thin Irish kid from San Francisco. I could see that he was getting the best of Casey so I got between them, and using a few fast punches, I knocked the bully to the floor.

He soon found out that he had much stronger competition than the ones he had been picking on. I let him get to his feet and after that, the bullying stopped, the fighting was over. What boxing experience I had before going to boot camp paid off.

In the afternoon we watched black-and-white World War II movies of everything from how to recognize enemy war ships by their silhouettes to abandon ship movies that showed us how to survive at sea. We were given lectures on what to do and what not to do if we had to spend hours in the water or possibly weeks in the salty ocean in a lifeboat, exposed to foul weather or blistering sun with only the clothes on our back to help keep us afloat, and for shelter.

The big dark room where the movies were shown had no air conditioner and was sweltering hot. It was a hard place to stay awake. It was really the wrong time of day to show movies, considering reveille was at dawn, and that after morning chow, we spent our time on the grinder doing drills. Thiess and I lucked out and took seats in a back row not far from the door where the ventilation was better.

Some of the sailors in front of us would start to nod off when there was a pause in the movie while the guy running the projector was changing reels. It was funny watching some of the guys in the rows of seats ahead of us when they quickly jerked their heads up when being nudged by the sailor sitting next to them. It was funnier yet when Thiess, who was sitting next to me, started popping some of them with a rubber band. Thiess, the "Joker," would quietly take a long rubber band from his dungaree pocket and wait for a couple of the sailors sitting in front of us to nod off. He would quickly snap them on the ear or back of the head with the big rubber band. It was tough keeping our laughter in when they got snapped and turned around whispering obscenities and threats to whoever was doing this to them.

After realizing he would be in big trouble if he started a fight, he quit snapping the recruits and slipped the big rubber band back in his dungaree pocket.

How Thiess managed to finish Boot Camp without getting in serious trouble or kicked out of the Navy, I'll never know.

I graduated in August 1954 with grades high enough to qualify me for Machinist Mate School, a basic steam propulsion school located on the shore of Lake Michigan.

First Leave

I went home on leave for thirty days and then took a Trailways bus out of Burns. This time it would be to the train station in Ontario where I took the east-bound train to Chicago. From there I would take the northbound train that runs between Chicago and Milwaukee to the school in Waukegan, a Navy town, about thirty-five miles north of Chicago on the shore of Lake Michigan.

The train ride from Ontario to Chicago went fast, a lot faster than I expected. It was different than the ride from Portland to San Diego. There weren't nearly as many stops at small town stations.

Once we got over the Rockies, over the Continental Divide or "Great Divide" as it is sometimes called, the long passenger train really picked up speed traveling down the long slope. The big train was moving' out, with the rapid clicking of the rails and the telephone poles flashing by like close-together fence posts. We were flyin' low, streaming down the long slope. I know the train was probably doing a hundred miles an hour or more when we came out on the flat.

Most of the servicemen like me were traveling in the coach or "chair" car. Traveling coach wasn't all that bad. The rough part was trying to reach a comfortable position to go to sleep in the big reclining seats, which were bigger but not much better than Greyhound bus seats. I enjoyed walking from car to car visiting with people from different states and different countries. The food was good and I especially enjoyed the beer.

When we pulled into Chicago it was early in the afternoon, and after about a twenty-minute wait, I boarded the train that runs between Chicago and Milwaukee going north and got off at the Naval Training Center in Waukegan.

I checked in with the main office and then was directed to the office in charge of the Machinist Mates School where I filled out all the necessary paperwork. After going to the barracks and getting everything squared away in my locker, I, along with several others that were to be in my class, headed for the chow hall. It didn't take long to get acquainted with the guys in my class.

My weekend trips to Chicago and Milwaukee were a lot of fun, especially for those of us who enjoyed trying different kinds of beer as well as varieties of good cheese. The bars had free snacks of popcorn, peanuts and pretzels and in some bars they had like a little buffet table where you could get small portions of kraut and wieners. All this good stuff made us really enjoy the different flavors of the draught and bottled beer. The bars in the towns of Kenosha and Racine just north of Waukegan were also good places to enjoy this kind of recreation, with juke boxes blaring out some of the latest songs of that time, like "This Old House," "Sincerely," "Skokian," "She Boom She Boom," and "Hearts Made of Stone."

The train rides north were scenic, especially after the first snow. The pheasant hunters with their dogs walking along the hunting trails were quite colorful. The passing scenes had a picturesque array of mixed colors; some trees that had not lost all their leaves were dusted with the light snow like pictures on a calendar.

On one of those return trips coming back from Milwaukee around midnight, we witnessed the northbound train plow into a '51 Ford convertible that had stalled on the tracks. There were three sailors who had been trying to push the stalled car off the tracks. They jumped clear just in time as the big engine struck the car, sending a rooster tail of sparks coming out the side and shooting up over the cab of the big engine, making the car look like a big piece of metal being shoved against a giant grinding wheel. The car was dragged several hundred feet before it came to rest in a ball of burned and crumpled metal and broken glass.

I'll never forget those train rides.

We had no duties on the base on weekends and most of the guys in my class spent their time in Chicago or Milwaukee. The few times I took liberty in Milwaukee I enjoyed stopping at the candy and cheese shops between my bar stops, but we all really enjoyed going to the Loop, a section in downtown Chicago. The Loop with all of its glittering and flashing lights was quite the place. There were all kinds of things to do and see.

There were arcades crammed with challenging games of every kind and up and down both sides of the street; there was everything from peep shows to movies and lots of little food shops—you could get everything from fish and chips to hamburgers and pizza by the slice to corn on the cob. It was the first place I'd been since we were living in Denver with so many glittering and flashing lights. It was a real fun place to be.

The train rides to different parts of the city were quite an experience. Just riding the elevated train, the "El," was fun. I was really amazed at the number of trains going in all directions crisscrossing the city. You would board the train above ground and within minutes you might be going through a long tunnel and all of a sudden pop up on the big steel bridge struc-tures with the sun streaming in through the windows as you looked down at the traffic and pedestrians be-low.

There were famous places to see in the city like the Biograph Theater where the F.B.I gunned down the famous bank robber John Dillinger as he came out of the theater with the Lady in Red. One of the most interesting places I visited was the famous Museum of Science and Industry. It had a display of historic planes hanging from the ceiling and the famous U-505, a German U-boat that was captured near the end of World War II. These places were especially interesting to a country boy like me.

Fuller, a member of my class, wanted me to meet his fiancée, a girl he had met in Chicago and planned to marry after he graduated. She was still living with her family in one of the tall apartment buildings. Their small apartment was several stories above the main floor.

I met the nice family, his fiancée, her mother and dad, and her little sister. Fuller and his future in-laws showed me the different views of that part of the city from the apartment and views of the street below. The small, two-bedroom apartment included a small porch where they kept their broom and mop. The porch had a clothesline that was rigged up with pulleys that ran a line over the alley to the other big apartment building next door. As you looked over the porch rail toward the alley below, you could see clotheslines running out from the apartment porches below.

His fiancée's little five-year-old sister had just recently added a new word to her vocabulary: "obvious." It was so much fun talking to her. When we told her how pretty her dolls were, and when I mentioned how pretty she looked in her new dress, she looked up at me and said, "It's so obvious." We all laughed. She really felt grown up using that word. She managed to use the word "obvious" in everything we talked about, any question we asked or statement we made.

One weekend a couple of us checked out Calumet City. "Cal City" is about thirty miles from Chicago. We had heard about the nice clubs and pretty women there, but were disappointed after checking it out. There were pretty women all right, but they were coming and going with men in their fifties and sixties, old guys with lots of money, some driving new Cadillacs and some getting out of limousines. After checking out a few of the bars we went back to Chicago to finish our liberty time before catching a train back to Waukegan.

Machinist Mate school went by pretty fast, but I did get to go places I'd never thought of going. One long weekend I was invited by one of the sailors in the class, Neptune (an appropriate name for a sailor), to spend the weekend at his home in Ohio, near Cuyahoga Falls. The Neptune family lived between Cuyahoga and Akron. They were having a family party, and invited me to join the fun. He said the party would last all weekend and spill over into Monday afternoon if I wanted to stay, and he said that I would have a chance to meet and dance with some pretty, nice girls who were going to be there. They were friends of the family. We rode down on a Greyhound bus; the weather was bad and it rained all the way. We became really irritated by the time we got to Akron.

There was a black man sitting in the seat behind us chewing bubblegum and popping his gum the whole trip. It sounded like someone clapping his or her hands together each time he popped. The bus was crowded and no one had the choice of moving to an-other seat. The loud popping made it almost impossi-ble to think straight and carry on a conversation. We were really glad to reach our destination and get off that bus before one of us blew his stack.

I enjoyed partying and the company of some of the girls my age that were there. After a few drinks, Neptune rolled out of the back bedroom on roller skates and started rolling from room to room. If you needed another beer or a snack, he rolled to the kitchen and got it for you. To this day, I don't see how he kept from falling and breaking his neck with all the beer he drank while skating between the rooms. The party was fun and most of us drank and danced until almost dawn.

Christmas in Chicago

I spent Christmas Eve in Chicago with three of my classmates. The four of us were like many others in the school who lived in states too far away to go home over the Christmas holidays. Having spent time there before, we knew that like the other cities in and around Chicago, the people treated servicemen in uniform very well. We decided to go to the city and enjoy walking around the Loop where we could tour the shops and shows. Chicago was every bit as cold if not colder than Cripple Creek. In just minutes, the freezing wind blowing off of Lake Michigan would numb your face and hands. You definitely had to stay on the move if you were doing anything outside.

We had just walked out of one of arcades and had started down the street when two pretty girls pulled up in a nice four-door Cadillac. One of the girls on the passenger side rolled down the window and started talking to us. They said they had plenty of food left over from Christmas dinner and asked us if we would like to go to their home and share some of it with them. They said where they lived wasn't from where we were.

The one driving the car asked me if I had a driver's license and I told her I had an Oregon license. She asked me to drive and said she would give me directions on how to get to where we were going. The three guys I was with got in the back of the warm car and, I took the wheel, and we started down the busy street. I think she wished she had never asked me to drive when she suddenly told me to change lanes and we were almost sideswiped. The girls were sisters, obviously from a wealthy family. What a lucky break.

After about a twenty-minute drive through heavy traffic, we arrived at their nice big house. When we went inside, we were introduced to the other members of the family, some of the women working in the kitchen and others straightening up the furniture in the big dining room. Some were still at the table having dessert where a big platter holding part of a turkey set with all the trimmings was sitting nearby. After a brief visit, the family asked us to follow them to the game room downstairs. The big room had a full-sized dining table already loaded with a second big turkey and all the trimmings. There was also a pool table, ping-pong table, and a gambling table with a deck of cards and poker chips. None of the four of us had ever seen anything like it in a private home.

One of the girls carved the turkey while the other one filled our plates and wine glasses. What a nice family; what a nice place to enjoy a Christmas dinner. After stuffing ourselves with all the good food and wine and playing several games of pool it was soon close to midnight. We gave them our address and they drove us back to the Loop. We thanked the girls over and over again for the good dinner and fun time we had as we got out of the car and started down the street.

We pooled our money to see if we could stay the night in one of the second-class hotels instead of going back to the training center. We already had our return train tickets and managed to come up with fifteen dollars between the four of us. We asked a man who was working in a pizza shop where we could find a cheap room for the night and were directed to an old hotel. I think it was the oldest one in Chicago, and the rooms we rented were run-down but clean.

The old building had settled on one side and the stairs between floors were on such a slant that you leaned against the wall when you used them. And the little screened-in elevator was not much better. The clerk or manager took us up to up to the fifth floor and showed us a room that was available. The little room had the light by the door with a string light switch and the transom above the door was missing the glass and had a piece of barbed wire nailed across the length of it. The room had four small beds with a closet-size bathroom at the end of the hall. The bathroom was shared with the people staying in the two other rooms.

The setup was more like a flophouse, but what more could we expect for twelve dollars?

It was about one thirty when we finally got to bed. The cotton quilts that were on the beds were worn out; they had been washed so many times all the cot-ton was in lumps and the material that held it together was like cheesecloth. They were about as warm as sleeping under a window screen. The hot water radiator in the hall kept the room at about sixty degrees.

I woke up about three in the morning with my legs cramping and feeling sick. I jumped out of bed and it was all I could do to make it down the hall to the restroom before throwing up.

There were no paper towels and only half of a roll of toilet paper, not nearly enough to clean up the mess I had made. I made it to the broken down sink, cupped some water in my hand and rinsed my mouth out and stumbled back to my room to lie back down.

Enough said for the night in the miserable hotel, so much for "Christmas in Chicago."

When it came time to graduate I was pretty proud of the shoulder patch that stood out on my dress blues; I was now officially a " Mate Fireman Apprentice." The rate patch was a white three-blade propeller with two bright red fireman apprentice stripes beneath it. Those who graduated with the highest marks were the first ones to pick the ship of their choice on either coast. There I was again, not taking my schooling too seriously.

When it came my time to choose a ship on the west coast, all the fighting ships were already chosen, and the rest of us who were middle-of-the-class graduates or near the bottom of the list had nothing like destroyers or carriers to pick from, especially on the west coast. I ended up choosing the USS *Zelima* AF-49, an auxiliary, or "refrigerator" ship with its homeport in Oakland, California. I planned on transferring to a destroyer when I made third class.

Visiting Family Pre-Deployment

The train ride from Chicago to Ontario, Oregon wasn't the most pleasant of train rides; not saying the train itself was bad. The streamliner, the City of Portland, was one classy train. The big silver engine had big red roses with bright green leaves painted on its sides and was no doubt the pride of Portland. The coach cars were full of sailors from schools and those or going home or returning from leave. I had an inside seat near the aisle. The seats weren't all that comfortable for real long trips. That night I was painfully awakened when someone kicked my left foot. I bolted upright in agonizing pain, and by the time I realized what had happened, the person who kicked me was just going through the door into the next car. It was one of the porters. I'm sure he had been stumbling over several of the drunken servicemen sprawled throughout the cars and decided to get revenge.

After my leave with the family in Burns I was on the road again, headed this time for Treasure Island where I would get the necessary shots and travel papers before flying to Japan. I was scheduled to fly out of

Travis Air Base on a MATS (Military Air Transport Service) plane from the military airbase north of San Francisco in Fairfield.

I was up before dawn along with dozens of others who were on their way to Travis. We stood in the shivering cold behind our billet numbers in our dress blues and P coats with our sea bags on the ground be-side us waiting for our number and name to be called, and our orders checked out, before boarding the big gray bus that took us to the airport. I'll never forget that miserable morning when the petty officer pur-posely screwed up when he called out our last names. I cracked up laughing when he called out my name, calling me James "Cheekmore."

After our short ride we went through the gate, and before I realized it, we were alongside a huge four-engine propeller plane with three tail fins, what I later found was a Super Constellation made by Lockheed Aircraft.

As I stood there waiting to board, watching several of the big Constellations loading cargo and passengers, I started thinking back to my first and only plane ride I had ever had–the plane ride in Burns I had gone up in a Piper Cub flown by Claude Funk, the guy I worked for in Burns at the body and fender shop.

That first plane ride had scared the living hell out of me. I was doing fine until Claude started showing off, doing stunts with the little plane. He started making steep dives and loops that almost caused me almost to soil my pants. I couldn't wait to get on solid ground. I was sure glad we made a safe landing.

As I went up the boarding steps of the Constellation, I realized that for a small town guy like me, this was a real experience. I was on my way across the Pacific, flying to Japan. This was really something, and I was getting to do this after I had been in the Navy only a little over six months. My orders were to fly to Hickam Field in Pearl Harbor and from there to Tokyo. This was just the beginning of the things I would get to do and the foreign countries I would visit through-out the time I would spend in the Navy. After finishing boot camp or finishing one of the schools, many of the sailors never got to leave the states.

As we boarded, we were greeted by a first class petty officer, our flight attendant, who took a copy of our orders and instructed us where we were to sit. The big plane was rigged with hammock-like, flat-strapped nets with restraining belts. The flat-strapped nets could be used as passenger seats as well as for strapping down cargo. The only real seats were for the flight crew and pilot and co-pilot. There were only about a dozen of us and we were able to choose window seats if we so desired. To most of us, this was a new experience.

We were given Mae West life jackets (named after a buxom movie star of the time) and instructed what to do if the plane were forced to land in the water, or to "ditch" as they called it. We had sack lunches, plenty of hot coffee, and a candy bar for dessert. I munched on a Milky Way bar as we taxied to the end of the runway. The big plane began to vibrate as the pilot revved up the big radial engines, waiting for the signal from the tower for takeoff.

All at once we were thrust against the backs of our seats as the pilot opened the throttle sending the plane charging down the runway. As I gazed out the window the tarmac beneath us, the big plane moved faster and faster. The planes standing by waiting for takeoff were soon behind and far below. We were on our way. I was flying to Japan.

As we flew west, the clock was moving backward; it was still fairly early in the day when we made a smooth landing at Hickam Field at Pearl Harbor. After picking up a few passengers and some cargo, we topped off with fuel, and the big Constellation was once again taxiing out to the runway for takeoff. This time the scenery and the surrounding buildings were quite different. As we taxied down the runway, I noticed some of the larger buildings coming into view still showed traces of the attack on Pearl Harbor. One two-story building had strafe marks all the way up the side, almost up to the roof-line where a Jap Zero had flown in emptying its guns, strafing the buildings and people below.

As we sat on the runway waiting for the tower to clear us for take off, I realized that I had only seen this kind of scene in the newsreels they showed at the movies. As we became airborne, the scenery below was equally as moving. As we circled the field, I stared down at the clear bright blue ocean lapping at the white sandy beaches with their tall palm trees swaying gently in the ocean breeze. I just couldn't believe I was getting to see all of this. Our next stop was the Midway Islands.

The Midway Islands, a possession of the United States, are about thirteen hundred miles northwest of Honolulu, halfway between Honolulu and Tokyo. The coral atoll is about six miles in diameter. There are two islands. Sand Island on the western side is almost two miles long and almost a mile wide.

The other, Eastern Island, is over one mile long and a half-mile wide. In 1941 it was considered a national defense area. In 1939, before the war, the U.S had started construction of a submarine base and an airbase on Eastern Island. When the war started on December 7, Japanese warships shelled the American in-stallations. After that, the Japanese made random air attacks and on June 7 and June 8, 1942 an air battle took place between Japanese and American large aircraft carriers about 700 miles off Midway, which ended in a severe loss and defeat for Japan. It turned out to be one of the decisive naval battles and turning points of the war, the turning of the tide.

The brief stop at Midway was only for an hour and we weren't allowed to leave the plane. We were again brought food in paper sacks, a couple of sandwiches and some fruit, enough to hold us the rest of the trip.

With another smooth take off, we were soon out of sight of land. There was nothing down there but miles and miles of water. I sat there looking down and started to get homesick, thinking about home, thinking about Mother and the family, wondering how everyone was getting along, wondering how Tom was doing in school and thinking about how talented he was playing solos on his saxophone. Everyone in Burns always looked forward to hearing him play. He was not only a saxophone player, but also very good at playing vibes or xylophone and bass fiddles, both electric and acoustic. He eventually made music his profession. His life was music. He played up until his death in 2003.

The sun was low in the sky, as the plane banked toward Tokyo and as I looked out, I could see Mt. Fuji off in the distance, partly covered in snow. It wouldn't be long until we would be landing in Tokyo and I would be setting foot in another place I'd never dreamed of going.

5 UP IN THE AIR AND UNDER THE SEA

When the plane came to its final stop it was surrounded with transfer trucks for off-loading the cargo, and buses for the passengers who were on their way to ships or bases where they were stationed. I boarded a bus that would take me to Atsugi airport; I was to be flown from there to the Itazuki airport (now renamed the Fukuoka airport). My flight from there would be in a PBY, a seaplane that would land in Sasebo Harbor.

During my bus trip from Tokyo to Atsugi, we passed large sections that were littered with piles of rubble, which were the result of our bombs dropped there in World War II. On the outskirts of Tokyo, the bus trip was delayed about twenty minutes while a crew of Japanese laborers pried out a big water line that crossed the road.

Dressed in their heavy winter clothes, their sandals sitting by the roadside, they were only wearing socks like I had never seen before that were more like gloves with a pocket sewn separate for the big toe. There were four men standing on a pole with one end of the pole placed underneath the pipe, like a huge pry bar. When their foreman shouted the order, they would all jump up at the same time, landing hard when they came down, causing it to inch its way out of the ground. There were also two men on each side of the pipe moving a bar under it to keep it from slipping back down. They would pull up on the long bar that ran under the pipe just as the eight stocking-footed men landed on the pole. They were very coordinated. As we pulled away, they were dragging the pipe away and backfilling the ditch.

When we drove into the Atsugi airport I was in for another surprise; after getting our clearance and going through the gate, the bus drove straight out to one of the several PBY seaplanes that were parked near the runway. Six of us boarded a twin-engine plane. The door was closed just as the big radial engines came alive with the co-pilot handing us our life jackets and shouting our instructions over the roar of the engines as we strapped ourselves into our seats. It was a short but bumpy flight from Atsugi to Itazuki. We made a brief stop at Itazuki, took aboard two more passengers and some mail sacks, and took off for Sasebo Harbor. I was a little uneasy as we dropped down near the water. There were whitecaps kicked up by the strong winds, which made for a rough landing. As soon as we touched down, the plane skipped over the waves and settled down to a slow cruise in the harbor. When the pilot cut his engines and we came to a stop, we were only bobbing in the choppy waves a few minutes before a boat came alongside to pick up the mail and passengers. The boat made several stops at the various Navy ships in the harbor, letting off sailors and exchanging mail sacks. We didn't stop to let me board the *Zelima* AF49

(Auxiliary Freighter). I would board her the next day. After checking in at the barracks where my orders were checked and I would spend the night, I took liberty with several others and took a stroll downtown. I had very little money to spend, and after browsing in a few stores and downing a few bottles of Asahi, I returned to the barracks.

The Zipp'n Z of the South China Sea

My service time aboard the USS *Zelima* AF49 was any¬thing but pleasant; in fact, it was the worst tour of duty I spent in the Navy. The short time I spent was miserable. It was surely not the kind of ship I wanted to spend my service time on. The USS *Zelima* was commissioned the second of March, 1945 in Oakland, California. Her original name was the *Golden Rocket*. "Golden Rocket"–what a joke. With a speed of sixteen to nineteen knots, she was anything but a rocket of any kind. She became a Navy ship when turned over to the Navy and converted to a stores ship, or refrigerator ship, for transporting frozen foods to the ships near Japan and the coast of Korea. She was renamed the *Zelima* in 1946. I have no idea what the name *Zelima* meant, but I do know it had nothing to do with speed. She saw constant duty off the coast of Korea as a food supply ship in the Seventh Fleet. After the Korean War in 1953, she continued supplying the ships off the Korean coast as well as some army and marine units ashore and air squadrons flying in and out of the small islands near the Korean Peninsula.

After morning chow, those of us who were to be taken to the various ships that were in Sasebo Harbor stood in line as our papers or orders were checked. We were then escorted to the waiting whaleboats tied up alongside the dock facing the harbor. The open boat I boarded had about twelve men including myself, plus the coxswain and his line handler.

My ship, the *Zelima*, was the third stop. She looked bigger and bigger as we pulled in close. When we did come alongside, I could see the small floating dock with steep boarding steps going up the side to the main deck, which was a long way up. I hoisted my sea bag from the deck of the swaying whaleboat, set it on the small floating dock, and stepped from the rocking boat to the swaying dock.

I squared away my uniform, making sure everything was in order—cuffs buttoned, hat squared away, and tie straightened. I then shouldered my sea bag and climbed the steep boarding steps to the main deck where a Lieutenant J.G., the officer of the deck, greeted me. I lowered my sea bag to the deck, saluted, and asked permission to come aboard. I then handed him my orders and after he read them I was introduced to a first class petty officer who showed me to my quarters. I unloaded my sea bag into my locker and went back out on the main deck where a first class boatswains mate was waiting to take me on a tour of the ship.

I was shown where to find the fire extinguishers, fire axes, and lifeboats, and all the things that would be most important in case of fire or collision. He showed me the boiler room where I would be working. I was surprised at the sudden heat that blasted us when he opened the door and we stepped inside the compartment leading to the boiler room.

The temperature must have been a hundred and twenty degrees. The "uptakes" as they were called (a maize of steam lines wrapped with asbestos and stenciled with arrows pointing in different directions), came and went from the evaporators and the boiler at the lowest level. The catwalk leading down to the boiler and evaporators was a series of ladders and grates that were like small decks used in the maintenance of the steam lines and valves. The next morning at muster, I would get acquainted with the engine room gang who I would be working with.

It didn't take long to separate the good guys from the bad guys, and it didn't take long for me to decide to try to get off that ship as soon as I could. I would either try to go to another school or try to transfer to another ship. The first class petty officers that I worked under were quite overbearing. There was one in particular who none of us could stand. He was a real piece of work. It wasn't just his overbearing atti-tude, but also his looks. We were pretty grossed out when he came out of the shower to get dressed, seeing a dark human body so fully covered with shaggy black hair.

Back in the forties and fifties, there was a comic book called Black Hawk. It was about a special air force that was always battling demons and monsters. This gross first class sailor (MM1) reminded me of one monster in particular, a swamp creature called "The Heap." The shipmates I worked with cracked up when I told them who I thought he looked like.

Hardly anyone in the boiler room liked The Heap. He was a real smart-ass. When we gathered in the boiler room for our job assignments, he would say he had a special job for a special person and those of us who were the lowest in rank were given those "special," dirtiest jobs. We expected this; but coming from this man, Mr. Gross, The Heap, was another matter. His disgusting smile made us resent him all the more.

I can't say I didn't have fun in the different ports we visited before heading back to the States. I, along with other shipmates, enjoyed being peddled around in the three-wheeled bicycle rickshaws with their canvas covers and small Hibachi heaters on the floorboards as we toured the gift shops, nightclubs, geisha houses and bars. We enjoyed getting loaded, drinking at the EM Club (Enlisted Men's Club) where I went on stage and played bass fiddle with the Japanese band playing Western music.

One night at the EM Club, I got in a fight with one of the crewmembers who hit me in the head with a beer bottle before we were ushered out to liberty boat and taken back to the ship. We were lucky neither one of us got in trouble.

The geisha houses in Sasebo, Yokohama, and Yokosuka were especially fun. The busy crowded streets were hustling and bustling with people selling anything and everything they could to make a dollar, trying to get back on their feet after the war. The yen exchange was 360 yen for every American dollar, so our liberty money went much farther than it would have back in the States. The street vendors had just about anything you wanted in the way of souvenirs and trinkets from Japan and we could have our shoes resoled if we had time to wait.

Even if we didn't buy from them, they still treated us with politeness and respect. They bowed, thanking us politely, moving out of our way as we weaved between rickshaws and people pedaling small-wheeled bicycles carrying everything from big bundles of cherry wood to three or more cages crowded with live chickens and ducks lashed to a wide rack behind the seat.

The pretty geisha girls met us in front of the geisha houses, tiny girls in bright kimonos waving their colorful fans while talking to us in pidgin English as we left our shoes at the door and walked to our rooms.

It seemed like every time I came back aboard ship after being on liberty, I would end up with my pea coat pockets stuffed with funny cards describing the geisha houses and their geisha girls. The "Grade A" houses were supposed to be the places of choice. They were the ones we were advised to use by the officers aboard ship before we went on liberty. They were the cleaner ones where we would be less likely to contract any kind of venereal disease. They were the houses where we would receive "special attention."

Here is a sample from my collection of the cards:

> *Night Club U.S. Isezakicho Yokohama*
> *Number 1 Place*
> *Oasis of Your's Number 1 Hostesses*
> *of Course Sweetest Number 1 enjoyment*
> *Never had experienced.*
> *No Body Can Find-Out More Nice Place Than*
> *U.S Club in Yokohama*

One of the clubs in Yokohama that featured an "Exotic Skin Show" had a sexy, almost-nude dancer who came on stage with a large boa constrictor wrapped around her. It was part of her costume. When she danced to the front of the half-rounded stage where I was sitting at a small table sharing pitchers of draft beer with a couple crew members and some other sailors, a couple of them doused her with beer causing us to be quickly ushered out of the place. The snake dropped to the floor and slithered for darkness at the back of the stage. We were very lucky there were no shore patrol or Japanese police nearby or we would have been in big trouble.

We left the bars and staggered down the busy street as cards were being handed to us in between bars. Another card I kept from downtown Yokosuka reads:

> *...Secret Where? What?*
> *At Geisha House Tagoto*
> *Please come up any-way and you will find*
> *Every-thing Such as Secret Things Done by*
> *wonderful women or Bath Room Furnished*

*in European Style and so Forth ....No
Other Place in Yokosuka Where We can
Serve You Like We Can Do.*

The toilets in the houses dumped into holding tanks. The water was drained off, and in some places, the sewer water ran into the gutter and the heavy stuff, or solid wastes, was gathered in tanks. The solid stuff was dipped out into carts holding tanks called "Honey Buckets." These were spread on vegetable gardens for fertilizer. Needless to say, we weren't too interested in chomping on any of the huge carrots or other (healthy looking) produce.

Most of my time on the way back from Yokosuka to Oakland was spent cleaning bilges and chipping rust in the "coffer dams," which are those small, watertight, inner bottom compartments sealed off by vertical manhole covers that were bolted between the small compartments. In some of the coffer dams there was up to a half inch of rust that had to be chipped off, put into buckets, and passed through the crawl spaces from compartment to compartment. They were then painted with strong, anti-rust vinyl paint. I think of having to breathe all that dust from the rust and strong, almost-overwhelming fumes from the acrid vinyl paint and wonder how our lungs held up when we were only wearing goggles and small dust masks with small low-pressure air hoses connected to them. Half the time they were getting tangled up with the drop cord for the light as we passed the rust buckets back and forth.

On our trip back from Yokohama we replenished several destroyers, emptying out the holds of the old refrigerator ship. It was hard to believe the amount of food that was transferred from the Zelima to the waiting ships. The sea was too rough for the destroyers and other ships to come alongside, even with the big hemp bumpers or "camels" as they were called (big hemp cushions slung over the sides of ships to keep them from banging the side of another ship or the dock while they were in port).

Trying to off-load that close in such heavy seas could cause serious damage to one or the other, or both ships, so the food had to first be off-loaded into small M boats to be taken out to the waiting destroyers. Some of the sailors helping with the off-loading of supplies made up a long sign out of butcher paper that they had taped to the side of the ship, out of the way of the cargo boom. The sign said. "You can whip our cream, but you can't beat our meat."

It was a sight to see the destroyers getting lined up to take on their supply of groceries. First you would see one bob up out of a deep trough, sometimes going completely out of sight as the M boat steamed out toward it. After receiving its supplies, it would drop down in the stern with its smoke stacks belching out black smoke, white water roiling up behind it as it quickly got underway. Soon it would be off in the distance with another one steaming up to take its place to get its supply of food.

On our return trip, my two main jobs or "watches" were the stern watch, in which I would report any vessel I thought was following too close and report every hour on the hour to the sailor answering the calls on the bridge the condition of the stern light. I would call in from my station and report, "Stern light burning bright."

My other job was below decks in the boiler room checking the propeller shaft bearings in the narrow tunnel that ran between the bulkheads, making sure that the bearings were getting proper lubrication and not over-heating. The big pedestal bearings supported the propeller shaft as it extended back toward the stern after leaving the big gearbox, which was powered by the steam turbine. They were evenly spaced clear out to where the shaft extended through the hull. The "shaft alley" watch, as it was properly called, wasn't that much fun either. Like I said, I could hardly wait to get into port so I could try to go to another school or somehow get a transfer.

San Francisco Bay

Our arrival into San Francisco Bay was early in the morning and the bay was blanketed in heavy fog. Along with many others, the vibration of the ship awakened me; the ship was backing down from a near miss of running aground on Alcatraz Island.

The harbor pilot that had come aboard earlier outside the Golden Gate had miscalculated the ship's position after going under the bridge and was headed too far to the east in the channel. After getting our bearing straight, we slowly steamed our way to Treasure Island, stayed there for a week, and then went from there to the port of Oakland.

In the summer months of 1955, the Bay Area was quite a bustling place. There were all different branch-es of military men filling the bars, nightclubs and plac-es of entertainment. The buses were crowded and there seemed to be no end to the sailors, soldiers and marines that crowded the sidewalks and street corners, people taking buses in all directions.

I had only been as far north as San Rafael on liberty and a couple of weeks after that visit when I had a four-day weekend, I was invited to go north to Siletz, Oregon. Siletz was the home of a sailor off the Zelima. I had talked with him the previous weekend in Del Monty's bar in San Francisco. He said that he was going to hitchhike up there and back. He was a Siletz Indian and all his family lived in Siletz. I decided to go just to get away from all the crowds.

Hitchhiking wasn't really my kind of travel, but since we both had very little money, it was the fastest way to get there and since we were in uniform we had no trouble catching a ride. Siletz is about fifteen or twenty miles northeast of Newport. I hadn't been that far north since my train ride down from Portland when I enlisted.

The visit was brief but fun. The worst part of the trip was coming back. We thought that we would make better time if we went inland instead of coming back down the coast–what a mistake. We got a ride to Mt. Shasta. We arrived about one o'clock in the morning and stood about two hours in the cold, neither one of us wearing any kind of coat. We were freezing cold when we finally got a ride and really appreciated our bunks when we got back to the ship.

The times I went ashore after that were spent in and around San Francisco, in the shops and bars. When The Heap saw some of his crew getting ready to go on liberty, he knew that a couple of us wanted to punch him out, so he would always manage to take a different bus to stay clear of us. He pretty much knew what would happen to him if we were drinking in the same bar.

I was anxious to get off of the *Zelima*. I had resolved to try my best to put in for another school if any were available. I was determined to get off the ship before it got underway again.

The morning after we returned from Siletz, we tied up a hardhat diver who was lowered over the side to check out the bottom of the hull to see if any damage had been done to the hull, and to check out the propeller after our near-miss of running aground on Alca-traz. He was a boatswain's mate and one of the crew-members. I stood watching the bubbles boiling up from his diving helmet as he searched for damage to the bottom of the ship. When he finally came back aboard and changed into his dungarees, I talked to him about my chances of going to deep-sea diving school in Bayonne, New Jersey.

After talking to him about the training he had gone through I was pretty sure I could pass the test. Then he told me about the worst parts, like having to put bolts and air hose fittings together on a metal plate simulating the hull of a sunken ship without any kind of light in thirty feet of soupy mud. He told also of the scary experience of having the diving suit purposely inflated causing him to go upside down in the dark muck.

For years as a kid playing in the irrigation ditches in Arizona, I had thought how great it would be to be able to explore the ocean floor.

There were two schools I put in for that late summer of '55. My first choice was deep-sea diving school and the second was submarine school. The diving school was full up, but they needed more volunteers for submarines. I lucked out and not long after passing the test in the decompression chamber at Hunters Point, I went to submarine training school in Groton, Connecticut.

I went through the test in the decompression chamber at Hunters Point with another crewmember, a first class petty officer named Jeske. Jeske, a first class machinist's mate, was one of the men in charge of the boiler room. He was one of the petty officers that kept me and the other "bilge rats" I worked with busy in and around the boiler room. I passed the test but Jeske failed it. I didn't know it, but I would meet up with Jeske later on in New London.

The decompression capsules were pressure tanks like those seen on butane trucks but much bigger. There were two tanks connected together, a large one and a smaller one that was used for those who could not take the pressure in the main chamber. Inside the main chamber, there were benches lining both sides for those taking the test. The test was to see if your ears and lungs could take the 33.3 psi (pounds per square inch) that you had to be able to withstand before going to submarine school in New London, Connecticut.

Those who had trouble breathing or earaches from the pressure would raise their hand, the pressure would be stopped, and they would go into the smaller tank where they could decompress. The pressure was slowly bled off before the door was opened. The 33.3 psi was the pressure you would be subjected to when you made your deepest descent in the escape training tank in New London, a towering round water tank that was designed for making mock escapes from a sunken submarine.

The last week aboard the Zelima was a busy one. There was a lot of activity all around the ship, going on both topside and below. The holds on the empty ship were bustling as the long line of refrigerator cars were being off-loaded onto the ship.

Cranes were swinging the pallets loaded with frozen food over and into the holds with a chain of crewmembers passing box after box to the sailors who put them in the big freezers and in stacks at the end of the chain. We had everything from frozen beef, chicken, and vegetables to ice cream. Some of these containers seemed to somehow get damaged and come open. Some crewmembers had their ice cream bowls ready and were gobbling down bowl after bowl while waiting for the crane to lower the next pallet of frozen food. There seemed to be no end to the line of refrigerator cars. I was told that it usually took a total of 110 cars to fill the ship.

New London, Connecticut

In sub school, we learned all the basics to familiarize us with submarines, and once we were assigned a boat, the real job of qualifying began. I worked hard and qualified in nine months. Being a machinist mate, I would be working in the engine room.

To qualify I had to trace out and make drawings of the different fuel lines, water lines and lube oil lines. After the nine months to qualify and earn my dolphins (a submarine specialist badge), I had completed twenty-four drawings of the different systems and qualified in all nine compartments, learning all the major things I was expected to handle in case I were in that compartment in an emergency.

All aboard were trained to know how to handle the basics in each compartment. The officer on duty or "skipper" in charge gave orders that were to be carried out throughout the submarine. The control room was directly below the conning tower (which contained the periscope), and it had the blow-and-vent manifolds for the ballast tanks, the main helm (there was another in the conning tower), the trim and drain manifold, and the bow-and-stern plane controls; and, of course, the "Christmas tree," a panel of red and green lights that showed the position of each ballast tank's vents and flood valves, and the snorkel intake and exhaust valves in the engine rooms.

When the officer in the control room calmly said "Green board" as soon as we submerged, the port and starboard lookouts came down through the conning tower hatch and took their positions at the big bow-and-stern plane wheels.

When you heard the words "green board," you knew all valves and vents were in their proper positions and the boat was safely on its way down. The Main Induction, a big 36-inch "mushroom" valve that supplied fresh air to all compartments when we were running on the surface was opened and shut by hand with a big swing handle in the crew's mess hall. It reminded me of an emergency brake handle on the old cars made in the thirties, only it was much bigger and swung down from the overhead.

When the diving claxon sounded with the order "Dive, dive," the big emergency brake-like handle was tripped and the big mushroom-type valve slammed shut. The big valve could turn out to be a real disaster if it got damaged and sprung a leak.

The blow-and-vent manifold in the control room was designed so that you could blow the main tanks even if the compartment was completely without light. The handles on the main tanks to be blown, in case the boat had to surface in an emergency, were made easy to find in the dark by the shape of their handles. The first tank that was usually blown was the "bow buoy-ancy" tank. The end of the valve handle was triangu-lar. The negative tank or "down express" as it was called was located in the center of the boat under the control room. It could be quickly flooded in an emer-gency dive or blown to help bring the boat to the sur-face. It had a square at the end of the blow valve han-dle, and the safety tank blow had a round handle.

The Escape Training Tank

One of the first tests that all new recruits had to pass when we started to qualify for sub duty was to go through the big escape training tank in Groton. If you failed to pass this most important test, you were bound for shore duty or assigned to a surface ship.

The big cylindrical tank had an elevator that ran up and down the front side of the tank. The elevator stopped at three different levels. It started at the top, eighteen feet below the water surface. The next stop on its way down was at fifty feet. The last stop was at one hundred. At each level there was a small "escape compartment" from which the person or persons qualifying would make their escape.

After the watertight door between the compartment and elevator was closed, the person making the escape was then instructed to open a valve to bleed water into the compartment until the pressure was equal to the pressure in the big tank.

When the pressures were equal on both sides of the door, the person or persons making the escape could then step out into the tank by ducking under the inner watertight door after taking a deep breath of the com-pressed air, which he then blew out after grabbing a toggle that that slid up the cable to the surface.

Everyone making the ascent is warned how dangerous and a real the possibility of death if you were to hold your breath on the way to the surface.

A good example of what can happen is when a deep-water fish has its entrails forced out of its mouth from the rapid decrease in water pressure while its quickly being reeled to the surface.

If anyone held his breath when he left the chamber, a diver watching your ascent from the surface would dive down and grab you by the arm as he lightly punched or pressed on your stomach to make sure you expelled your air as you made the ascent. Those divers and instructors were really in shape. It was unbelievable how they could dive in and swim down one hundred feet to make sure the man going through the test didn't hold his breath. During World War II and part of the Korean War, Momsen Lungs were used to aid persons escaping a sunken submarine. (Rear Admiral Charles B. Momsen had invented the breathing mask in his fifties.)

Hunter-Killer Subs

In March of 1953, the USS *Croaker* SS 246, built in the 1940s, had been taken out of the "moth ball fleet," for conversion. It was refitted with a snorkel, bigger, longer-lasting batteries, sophisticated radar and sonar equipment, and was then re-commissioned after her makeover and facelift on December 11, 1953.

She was reclassified and returned to duty with her new designation as a "hunter-killer submarine."

Not only did she have the latest in sonar equipment, but also a Bathythermograph, a graph that recorded the water temperatures at different depths. Water temperatures at different depths or layers have an ef-fect on the sonar of surface craft and submarines probing the depths trying to locate the enemy.

After the makeover in the Navy yard, the *Croaker* and other hunter-killer subs returned to their squadrons with things that helped to keep down noise to avoid detection both on the surface and when submerged. All the electric motors had rubber mounts that kept down vibrations that could be picked up, and both engine rooms had padded canvas mats covering their steel decks. The mats muffled any noise of anything dropped, like if someone dropped a wrench while working on one of the engines.

These improvements and others made throughout the boat made it almost impossible to be detected by ships on the surface or by other submarines. These improvements worked out very well while keeping track of Soviet submarines and surface craft during the Cold War with communist Russia.

The *Croaker*, equipped with a snorkel, was able to run its diesel engines at a depth of 58 feet. Not being a deep diving boat, the Croaker had a test depth of 412 feet. All submarines are tested for maximum depth and those that descend below their test depth are flirting with disaster. The pressure hulls are only designed to withstand a certain amount of pressure; going deeper and going beyond test depth could be a fatal point, a point of no return. This would cause them to implode as the pressure hull was crushed by the increasing water pressure.

All World War II submarines didn't have the same test depth and a lot of them sank trying to go deep to avoid the depth charges fired from enemy destroyers.

When the nuclear boat Nautilus came out, she was able to operate at 1000 feet–a depth deeper than any of the diesel-powered submarines. As a submarine increases its depth, the pressure hull elongates and shrinks; it actually stretches. It was noticeable aboard the Croaker when operating below 300 feet. We could tell by the unpainted lines that ran between the compartments above the watertight doors. These all returned to where they had been and were no longer noticeable as we got closer to the surface. The most noticeable noise was made when we made a fast ascent to the surface; parts of the superstructure made a slight noise as the hull expanded.

Submarine Duty

After sub school and basic diesel school I was assigned to, and qualified on, the USS *Croaker* SSK-246. I had no idea that my duty aboard the Croaker was going to take me to so many foreign places to see what I had only seen in geography books.

I reported aboard for duty the last of October, 1955. The boat was over-manned at the time because of some men waiting for transfers and some waiting to be discharged. My bunk was a mattress pad on an empty torpedo rack in the After Torpedo Room.

The skipper at that time was LCDR Dean L. Axene, a graduate from the U.S. Naval Academy in June of 1944. After his graduation, he reported for duty on board the USS *Parche*. He received the Bronze Star for action against the enemy when the Parche was on her sixth war patrol. The second in command was LTRB Mills, Jr. LT. Mills had come up through the ranks.

Dean L. Axene had enlisted in the Navy before World War II and eleven years later reported aboard the *Croaker* as XO (Executive Officer) in April 1955.

LCDR Axene had only been the skipper on the *Croaker* for about a few months before I came aboard.

Qualifying on the boat required a lot of hard work and concentration. Tracing out some of the systems and making drawings of them could be mind boggling, and like tests taken in public school, there were those who liked to pass them the easy way—by cheating, copying someone else's answers or drawings. Those who were caught cheating were returned to shore duty or transferred to surface ships. They certainly weren't allowed to become submarine sailors. Every phase of qualifying was treated seriously, even flushing the toilet, being able to follow the proper procedure of making sure all the valves and back-up valves were properly closed. If the valve on the main water line coming through the pressure hull wasn't closed good and tight, the water pressure, especially while submerged at 300 feet, could easily cause the commode to run over, flooding the head and other compartments with salt water and even cause an explosion if the water poured into the battery well and mixed with the battery acid (on battery-powered submarines).

I enjoyed the time I served aboard the Croaker. The first nine months, the months it took me to qualify and earn my dolphins, were rough; but I liked the challenge. I was, and still am, very proud to have served my country, put in my Navy time, and to have qualified with the enlisted men and officers aboard the *Croaker*.

I'm sure I could have done a lot better and got discharged with a higher rate if I had not gotten in trouble over the drunken fistfights I had while on liberty.

I'll never forget my shipmates and the crazy pranks we pulled on each other to keep down boredom. Some of us had nicknames. Mine was "Jed," after Jed Clampett of the Beverly Hillbillies.

The name got started when Jim Upton, a southerner from Bessemer, Alabama and I started joking using a southern drawl as we were having chow. Instead of Jim, he called me Jed. After that everybody started calling me Jed. Jim Upton had put in for U.D.T. school (Underwater Demolition Team) and when he left for the Caribbean, we all missed his humor.

♪

My first two months aboard the boat were not the best.

I was aboard a couple of weeks and had just settled in when I got in trouble with the shore patrol while on liberty in New London. I had been drinking with some of the crewmembers in Danny O'Shay's, a famous bar on Bank Street.

The Irish pub had quite a reputation. It had served drinks to many famous celebrities like Bob Hope and others of that time, and the walls were covered with autographed pictures.

By the time I got ready to head back to the submarine base, I was quite drunk. After leaving the bar, we walked up the street to the bus stop where we could have a sandwich and a cup of coffee while we waited for the bus to take us back across the bridge to the submarine base in Groton where the boat was tied up.

I was just entering the coffee shop when I bumped into a woman coming out with a cup of coffee in her hand, sending her sprawling to the sidewalk. In my drunken stupor, I was apologizing and trying to help her onto her feet when the manager of the coffee shop saw the commotion in front of the door and called the shore patrol who were a block down the street.

He told them that I had shoved her down and was a troublemaker. I tried to tell them that I accidently bumped into her. They put me in the patrol car and took me to the New London police station where I spent the night. It didn't take me long to sober up. I definitely didn't sleep that night.

Early the next morning while waiting for someone to take me back to the base, a civilian being held in a cell next to mine went berserk and broke the commode off the wall of his cell causing it to flood with water while he threw pieces of the broken commode out into the hall.

When the shore patrol came to take me back to the submarine base, they thought at first that the pile of porcelain had come from my cell. Thank God I didn't go on a rampage and do something that stupid. If I had charges like that against me, I would have done some serious time in a federal pen along with receiving a dishonorable discharge.

When I was returned to the base and went back aboard the boat, I went to captain's mast and ended up doing twenty-five days in the brig. The brig was on the base. Spending time in the marine brig was an ex-perience like no other. Most of us spent the biggest part of our time working outside. The daylight hours were spent cutting firewood for the stoves and fire-places on the base or doing janitorial work in the big Catholic Church. The janitorial job required only three people. I worked as a janitor for two days; the rest of the work I did was cutting wood.

Cleaning the church consisted of wiping the dust off the pews, cleaning the toilets, putting the Bibles in their proper places, and cleaning the priest's room where all the wine was kept. The two days I worked in the church were uneventful except for an inmate named Bartlett. (Bartlett was an alcoholic that was doing six months in the brig before being transferred to a federal prison to finish out his five-year sentence for going AWOL while in the Navy and forging false paperwork for a new I.D. to get in the U.S. Army.)

The guard marched us to the church and sat in a pew near the entrance while we did our work. When he was out of sight of the guard, Bartlett would be back in the priest's office gulping wine from the half-empty bottles (which never got him high). Being the alcoholic that he was, he was used to drinking much stronger stuff.

The woodcutting job was a job none of us would ever forget. The job itself wasn't so bad; it was good just being outside in the fresh air. As I think back, it was rough but hilarious. First off, we were all fitted with mismatched boots and ragged but warm clothes. The rubber boots were a real joke; not only were they mismatched and made by different boot makers, but they were different in size also. I marched out on the line with a ten-and-a-half on my left foot and a twelve on my right. My actual boot size was nine-and-a-half. The ragged, foul weather jackets were just as bad. Our worn attire was topped off with dingy worn-out sailor hats turned wrong side out and pulled down over our ears. We were a sight to see.

After dressing in this fine, degrading garb we were then marched down the main street for everyone to see, and then marched up into the woods, some of us bouncing along carrying with long, two-man crosscut saws springing up and down on our shoulders. After leaving the main road, the fun began. There were four guards each carrying a shotgun—two in the front, one on each side and two in the rear, one on each side. Before making our way toward the woods, we were given instructions on what to do if a prisoner was trying to break free and start running toward the woods. On the second command of "Halt," we were to drop our tools and fall flat on the ground to keep from being hit with flying buckshot. We got plenty of practice on the way to and from where we cutting wood. There was always a nice spot picked out for the drills, a spot where there was plenty of slush and mud.

While we were away sawing wood, the guards back at the brig were planning their next harassment. When we returned to the brig, we lined up in front of our cells at attention, our position for attention was standing with our heels together and our arms straight out in front of us folded at the elbows with our hands over lapping the forearms like a box, a position that was impossible to hold for any length of time without dropping one elbow or the other.

The guards made sure our elbows didn't sag at atten-tion by cracking you on the elbow with their nightsticks. After everyone was at attention, one of the guards unlocked the little cigarette cabinet on the wall and had those who smoked take a step forward, and while he was getting ready to pass out the cigarettes, another was checking out the cells of the ones who smoked. Just about every time he would come out of the cell holding up a cigarette butt (one that had been planted there, of course). He would then make the smoker stand in the center of the cellblock and smoke a cigarette with a bucket over his head.

The smokers were given their smoke break after a guard tore the cigarettes in half and walked down the line placing the broken half-cigarettes into the waiting open mouths. It was all I could do to keep from laughing as the first cigarette was lit and each smoker turned and let the next guy get a light off of his.

After their cigarettes were lit, they were allowed to sit on the floor in a circle and puff on their torn stubs. Sitting in the center of the circle was a square polished copper ashtray. The bottom of the ashtray was taped a Marine Corps insignia that almost covered the bottom, leaving a two-inch border where the smokers flicked their ashes and put out their stubs. They were very careful not to scorch the bright, paper Semper Fidelis emblem. It meant a lot of push-ups if it was damaged in any way.

I got quite a workout one morning when the guards were inspecting the cells. No sexy pictures were allowed in the cells. The only thing allowed in each cell other than your bunk was a Gideon Bible in which you were allowed one picture of a family member or girlfriend.

About three o'clock one morning, everyone jumped straight up in their bunks when the ear-splitting siren was turned on along with the crashing sound of a big wash pan bouncing off the cellblock walls. We all thought the place was on fire or some other catastrophe was happening.

After the siren was turned off, a guard screamed, "Everybody. On the line!" And as we all stood shivering in our underwear on the line in front of our cells, the guards were inspecting each cell. One of the guards came out of my cell holding a picture of Kathy Kennedy, one of the pretty girls I was dating in Holyoke, Massachusetts, a student of Amherst College.

He looked at me as he held up the picture and asked, "Who is this bitch?" I looked straight at him and answered back in a serious but humble voice, "A Marine, Sir." There was a long pause as the brig got real quiet; the cellmates were straining to keep from laughing. I thought for a minute that I was going to be backhanded. Instead I was ordered, "On the floor, give me eighty." Eighty push-ups. There was one thing for certain, we all came out in good physical shape, but none of us ever wanted to do brig time again, especially in a Marine brig.

After my brig time I reported back aboard the Croaker to start training to qualify in the different compartments. We operated as a school boat helping some of the sub school sailors get acquainted with what went on inside a submarine.

One afternoon I was in the After Engine Room explaining to students how the main air compressor worked, the one used for charging the air banks (the big flasks that were outside the pressure hull which were used for blowing ballast from the different groups of tanks). I was surprised when Jeske stepped into the compartment. He was equally surprised when he saw me. I could tell right away by the look he gave me when we shook hands that he was a little envious.

Here was the "bilge rat" that had worked for him on the Zelima. That bilge rat Creekmore was already finished with sub school and was now explaining to the students what seemed to me, the basics, about the multi-stage air compressor.

♪

As I've said, my first few months were definitely not my best. I drank too much while on liberty. On the afternoon of January the fifth, before we got underway for Caracas, I got into a fight with Lou Kimble, a shipmate I worked with in the engine room. We were both put on restriction and not allowed to go ashore.

We had dropped our whites off at the laundry and dry cleaners not far from where the boat was tied up at the State pier. It was afternoon when left the boat. We walked down the tracks from the state pier and dropped off our whites at the Chinese laundry and dry cleaners where we would pick them up that afternoon. We had plenty of time to pick them and walk the tracks back to the pier. We were getting underway for Caracas; our ETD (estimated time of departure) was 1600. We didn't have the maneuvering watch when we got underway, so after dropping off the laundry, we decided we would walk down Bank Street and stop at Danny O'Shay's pub to have a few beers and kill some time.

We were doing our usual fun things, slugging down beer and munching on snacks in between punching the jukebox and playing shuffleboard. Lou and I were the only ones off the Croaker in the bar. When we looked at the big clock above the bar, it dawned on us that in less than an hour they would be pulling in the gangway.

When we left the bar we were loaded and Lou was having a hard time walking a straight line. As we started down the tracks, I told him we were going to have to get a move on if we were going to make it back before the boat got underway. We would have made it in plenty of time if we hadn't got in a fight at the cleaners. As we were half walking, half-running, down the tracks, Lou stubbed his toe on a railroad tie causing him to fall flat on his face. When I helped him to his feet, there was blood running down his face where his mouth had hit the railroad track and split his lip, and broke off a big chip of his front tooth.

He wasn't really aware of the broken tooth until he looked into one of the big mirrors when we got into the dry cleaners. After paying for my laundry, I turned to Lou to get his ticket and some cash to pay for his laundry and uniforms. Looking in the mirror at his bloody mouth with the broken tooth, he turned as I approached him screaming, "Creekmore, you son of a bitch, you broke my tooth!" and he started swinging at me.

By the time the shore patrol got there, the dry cleaners was a mess. We cracked one of the big mirrors and knocked over several racks of uniforms. The manager had called the shore patrol. They already knew that the Croaker was getting underway at 1600 that afternoon and they had us get in the shore patrol wagon and let us off at the pier just as the Croaker was getting ready to pull away the gangway.

As soon as we got aboard, we were ordered below and Kimball went straight to the "Doc," First Class Pharmacist Mate Doc Mills, and had a temporary cap put on his tooth and his bloody lip bandaged. To this day I think the crewmembers thought I had punched him in the mouth and started the fight because of the swollen and skinned knuckle on my right hand.

We had all looked forward to leaving the freezing New England coast for "Springboard" operations (war games) in the warm waters of the Caribbean; but, the day before we tied up in Caracas the radioman on watch received a message that told us our ETD was early the morning of January the 16th and we were to proceed to Philadelphia to have a new battery installed. We had just made a 4000-mile trip for a weekend! Kimball and I were restricted from going ashore in Caracas. We would have to wait until we could go ashore on the different islands in the Caribbean.

♪

There isn't much to tell about the shore leave spent while in dry dock in the Navy Yard in Philadelphia. The big batteries were being replaced and new seals were put in various valves and pumps; the insides of ballast tanks were painted, flood-and-vent valves inspected and checked, and the overhaul of various electric motors and pumps were worked on in the big shop near the dry dock. A multiple-story barracks with a mess hall on the lower level was our living quarters while this and other important maintenance was being done.

Like the other enlisted men from the engine rooms, my job in the shop involved taking pumps and valves apart to replace worn parts and gaskets. The big shop, which was separate from the dry dock, turned out to be a nice warm place to work. We even had a nice big coffee pot brought down from the mess hall and had the luxury of plenty of coffee and big boxes of fresh chocolate and glazed doughnuts that were brought down from the mess early in the morning.

It seems like no matter what kind of people you eat with, there's always a glutton or someone that has to grab the biggest pieces of chicken or whatever he decides to wolf down, and has no consideration whatsoever for the others around him who might like some of the good stuff too.

Whitehouse, a third class engineman, was a regular pig when it came to chocolate-covered doughnuts. After watching him crowd in the coffee line and fill his paper plate, taking away more than half the chocolate doughnuts to wolf down with his coffee, Rainey Nichols and I, while trying to be halfway polite, were left out. When it came our turn for doughnuts, the chocolate ones would be gone. We were quite irritated but never said anything that might cause a row in the shop. Instead, Nichols and I came up with a plan to tend to the problem.

We were in the shop early sitting on crates having coffee when one of the cooks from the mess hall brought in the box of doughnuts. We each took two dough-nuts to have with our coffee before Whitehouse showed up. We were all set up to cure him of being such a pig at the coffeepot. As soon as we saw him walk in, we walked over with our paper plates to fill our cups and get another doughnut. As Nichols stood close behind, I exchanged the delicious looking dark chocolate doughnut on my plate with one in the box. Only the one I traded wasn't covered with chocolate–it was glazed all right, but not with heavy chocolate. Instead, it was covered with a shiny, sticky chocolate-colored gasket sealer called Permatex from a tube we had been using to seal the gaskets on some water pumps we had on the workbench.

Right off, Whitehouse loaded his paper plate with doughnuts. He filled his mug with coffee, sat down on a crate, blew on his coffee and crammed half a doughnut into his mouth before he realized he wasn't eating a delicious chocolate doughnut. Nichols and I and several others who knew what the joke was busted out laughing when he stuffed the doughnut in his mouth. The sticky Permatex stuck to the roof of his mouth and his fingers as he pulled out a partial plate holding three of his front teeth and part of the doughnut. The shop was in an uproar when they saw what had happened and were telling others as they lined up for coffee. Whitehouse worked half the morning trying to clean the Permatex off his partial plate and lower teeth.

It was a joke that would haunt him for several months. At sea, when he went to the crew's mess for a cup of coffee, someone would always ask, "Would you like a chocolate doughnut to go with that?" He never did find out for sure exactly who it was that doctored up his doughnut.

Some of the crewmembers like me who liked partying and dancing in bars with pretty girls, did have fun in the bars and eating places on the cobblestone streets in the city of the Liberty Bell. One of our favorite sandwich shops, or I should say a sandwich stand, was on the corner of Tasker and Passyunk, close to where we caught the bus back to the Yard. There I enjoyed some of the best steak sandwiches I have ever eaten.

The sandwiches came with different combinations. A dollar and a quarter bought my favorite, steak and hot pepper on a white roll with French fries. The big, grinder-sized sandwiches (we might call them "submarine sandwiches" now) that were cut in half were more than enough for one meal. I ate half on the bus trip back to the Yard, the other half I gave to one of the crew or had with coffee for breakfast.

Some of the bars and clubs had live music and were filled with pretty women, especially on Friday and Saturday nights. One bar the Croaker crew hung out at was called Biff's. It had a band that played Western music and jitterbug music, and plenty of girls who liked to dance and be escorted home for even more fun after the bar closed.

Only a couple of things that happened before I went on leave and went home, back to Burns, are worth mentioning.

Our government had sold a diesel-powered attack submarine to the Turkish Navy. The trip from the Philadelphia Navy Yard to their homeport in Turkey would be made running on the surface. We couldn't believe it when they begin to load the sub from bow to stern with everything from chest-type freezers, refrigerators, and all kinds of furniture and appliances, lashing them to the deck and covering them with tarps to protect them from the salt water spray on their way back to their home port. The Turkish officers were taking advantage of being able to buy good, dependable American goods.

I never saw it, but was told that the same officers had started to build a scaffold so they could perform an execution right there in the Navy Yard; but the Commander of Yard would in no way allow it and made them take it down.

Another incident worth noting was when a welder who was welding up some bracing underneath the grating just aft of the sail (conning tower) on one of the submarines in for repairs caught his coveralls on fire. A couple of shipmates and I had just come back from Philly and were inside the gate on our way to the barracks when it happened. One of the sailors I was with yelled, "My God, look at that!"

The welder with his coveralls on fire had just managed to climb out on deck and was wrestling with his coveralls, trying to get them off when another yard worker who wasn't thinking doused him with a bucket that containing diesel fuel and water. The dousing turned him into a flailing, flaming torch. We watched in amazement as he jumped over the side into the freezing water to put out the blaze. What a scene. Never in our lives had we ever seen anything like that except in the movies. One of the yard workers we talked to the next day said that the welder only suffered minor burns on his hands, "What luck!"

After the battery was replaced and other important maintenance taken care of, we would report back to Squadron Ten in New London and from there we would head out toward the Azores and on to the gateway to the Mediterranean, the Rock of Gibraltar. We were to join the other subs in Sixth Fleet (in Squadron Ten) in a show of force off the coast near the Gaza Strip, near the islands of Crete and Rhodes.

I felt so lucky. I was going to get to go ashore and see some historical places I had only seen in history books, some real interesting places in Europe.

♪

I went home on leave while the boat was still in the yard getting prepared to go to the Mediterranean. While I was home, I went to the Ford dealer, Mr.

Weeks, and picked out a car to drive back–a spotless, 1953 Ford Victoria coupe. It was a car that had belonged to Bud Garland's mother. Bud was a Burns high school classmate.

Mr. Weeks had sold my car after I enlisted and that money had gone as a down payment on the Victoria, leaving a balance of $700 to be paid out in monthly payments. I couldn't have been more pleased.

Mother and Tom decided to travel with me as far as Florida; from there I would go on to Philadelphia, and from there on to New London. We got a good start and I was able to spend a few days in Florida before heading for Philadelphia. On the way, I messed up big time at a service station in Colorado.

We stopped for an oil change and I left my nice, gold high school graduation ring in the service station washroom. I had left Tom and Mother in the car while I hurriedly went in to wash the grease off my hands after checking some things while the car was being lubed. I had pulled off my nice ring, which Mother had given me when I came home, and left it by the sink in the washroom. I hadn't been used to wearing it and didn't miss it until we were a couple hundred miles away when I was washing my hands at a rest stop.

I called the station, and was heartsick when they said they couldn't find it. The nice ring had a square ruby set in gold with two diagonal Mother of Pearl stripes and B.H.S in nice big gold letters. Mother felt bad too; she had worked so hard to get that ring for me.

Despite that turn of events, my short stay in Florida turned out to be an exciting five days. Lila's husband Phil who was an alligator hunter took us gator hunting one night in Needles Eye swamp. What a scary, I guess I should say, "adventure." It started out with a real thrill. Both Phil and Gene Creamer, the local gator hunter, were loaded on moonshine, some that Phil made and sold on the side to the locals in Wewahitchka.

It was pitch dark and the only light was coming from the bright lights on their hunting caps. The boat we were going out in was tied up to a small pier underneath a tangle of tall cypress trees. The swamp water was calm and the boat was dead still as we were handed our boat cushions and told to climb aboard.

I had just stepped down into the boat and was placing my floatable cushion where I was told to sit when Tom stepped aboard letting out a blood-curdling scream when a big black snake fell from one of the overhanging cypress trees, bounced off Tom's shoulder, and was slithering toward the bow of the boat when Phil grabbed a boat pole and flipped it into the water. Both Phil and Gene were laughing at the top of their lungs. Tom and I were both ready to back out but were talked into staying when they told us that those big black snakes were really harmless.

Creamer had been trying to catch a big 14-foot gator for weeks. He was running the boat with Phil holding a shotgun with a 12-gauge slug. They were both wearing caps with lights that shone out ahead of us as we made our way deeper into the swamp. We were surprised they found the beach where Creamer had a trap set.

He beached the boat under a big cypress to check out his bait. The big cypress had a three-foot length of chain holding what had once been a heavy barbed hook. The big, stainless hook had been in the center of the bait, which had been a couple of big chickens. The big hook was now just a straight bar without a barb. A huge alligator had taken the bait and straightened the hook.

As they shined their lights near the base of the tree, we could see the big tracks where the gator had walked to the tree, clamped his big jaws down over the bait and backed up, straightening out the hook. Creamer took measurements between the tracks of the front and back legs and a mark in the mud where the big gator's tail had been. He said the gator's length was close to fourteen feet.

As we cruised the swamp for the next hour, their bright lights were spotting the amber eyes of gators popping up and down like brown beer bottles floating in the water. We saw a lot of eyes but not a shot was fired. I guess they were too small to mess with.

Tom and I were amazed at how well Creamer knew that swamp, how he managed to weave in and out between the cypress and get us safely back to the dock.

After a few days visit, I headed back to the yard in Philadelphia. I was feeling great. I had not only gotten to visit the family, but I was taking a nice car back. I would now have my own car to take on dates. I would be able to cruise around in places like Middletown, Holyoak, and Springfield when I had weekend liberty.

I had no such luck. I didn't make it back to New London with my car. I tore up the front of it about two hundred feet from the entrance to the Holland Tunnel on the Jersey side of the Hudson. I was doing fine until the heavy traffic started bunchin' up.

I had never driven in that kind of a lane change. I was glancing at the "Squeeze" signs coming up to the tunnel, trying to steer clear of the cars crowding me on both sides where the highway was narrowing from four lanes to three going into the tunnel or "Hudson Tube." I was right on the tail of Nichols driving his practically new '55 red and white Ford convertible when he jammed on his brakes, coming to a dead stop in front of me. I hit my brakes, smoking the tires, dropping the front end down causing it to skid to a halt, wedging the front end under his back bumper.

We managed to get the cars unlocked with the help of some of the guys in the stalled traffic. I was one de-pressed guy. When we got the cars separated, my grill was wiped out, the hood was all bent up, and the radiator was pouring steaming anti-freeze all over the place where the fan blades had sliced into the front of it. Nichols and I couldn't believe what we were seeing. The front of my car was wiped out and his car only had minor scratches on the bumper.

I had my car towed to a body shop in Jersey City where it was to stay until my insurance gave the okay to for it to be fixed.

When Nichols and I got back to the Navy yard, most of the crew was already back in New London, and the boat was tied up at the state pier. A lot of the crew, like me, had been on leave until a few days before we were to get underway.

I never heard anymore about my car until I received a letter from the body shop with more bad news. When I had the car towed in, I was told that it would be okay for me to leave it there until I came back from the Mediterranean. I had told him that I would be gone at least three months. He said that would be fine, that it would be a month or more before he would have it fixed. He said the car would be safe and kept out of the weather under the big shed that was loaded with wrecks to be repaired where he had it parked when it was towed in. He said that he would notify me when to come and pick it up. I was in Weymouth, England when I got the letter telling me that the car had been repaired and the insurance had paid the bill, but said that he had repaired the car two weeks after I had left it and he had sold it for the price of the storage fee.

I was sick. I no longer had a car, and I had really been ripped off. Not only did I lose my car, the Bank of Portland was pressing me to pay off the balance of $700. I talked to a Legal Officer when we got back. I lost the car but didn't have to pay the balance.

Close Call in the North Atlantic

There were some life-threatening things that happened aboard the boat that could have been fatal for the boat and crew. One of these close calls happened while we were on patrol in the North Atlantic in the deep waters off the coast of Halifax, Nova Scotia. Our job was to monitor the traffic, both surface craft and submarines that came through our patrol area.

We were especially keeping track of a Soviet trawler that had been lowering a big round basket that looked like a crab pot and retracting it by a big winch on its stern, a winch that was much too big for even the biggest of fish or crab traps. Pictures of its actions were taken from a distance through the periscope. The big trawler, disguised as a drag boat for fishing, was monitoring sea traffic, using the basket to disguise the big sound head planted in its center. We were patrolling at 300 feet in when one of the sonar operators in the sound room picked up a fast traveling submarine headed south a couple hundred feet below us. Our code name was Miss Known.

Messages were sent from Miss Known to the unidentified contact asking for her identity while all the time monitoring her depth course. As the unidentified submarine increased her speed, we increased our speed to try to close the distance between us, but couldn't keep up.

When we received her code name and made contact, we found out she was the nuclear boat, the USS *Nautilus* SSN 571. She was at 500 feet and going deep.

Shortly after making contact, after recognizing the sub was one of ours; we started to plane up to snorkel depth to charge the then weak batteries. I was on duty in the After Engine Room when the order was given to surface. I was watching the gages and lights, and was especially concerned with the pressure gage that showed our depth. The needle on the gage was slowly going beyond the red mark, which showed our test depth.

I was standing by, ready to start the engines in the After Engine Room as soon we reached our snorkel depth of 58 feet when all of a sudden Nig Ross, a First Class Engineman who had served on the *Croaker* in World War II, came charging through the Engine Room with Dankievitch who was another First Class Engine Man known as an Auxiliary Man. His job was overseeing the maintenance of the auxiliary pumps and engines.

They quickly made their way toward the Maneuvering Room, tapping on the gages that showed the water pressure outside the pressure hull. The pressure gages in the different compartments throughout the submarine indicated we were going beyond the *Croaker*'s test depth of 412 feet. Ross and the First Class with him wanted to make sure that those on watch knew we were slowly going beyond our test depth.

Going beyond test depth was a real concern. Instead of making our normal ascent, we were at a steep angle, slowly descending, going deeper. We were headin' for the bottom. We were going down stern first.

Following the *Nautilus* for several minutes at full speed had drained the already low batteries, and when the electricians in the maneuvering room pulled the big breaker handles to put more power to the electric motors to turn the screws, they found they didn't have enough power. With the screws turning much too slow to give enough thrust to push the boat to the surface, we were continuing to slowly descend. Our submarine was on its way to the bottom.

We lucked out. Thank God, the air banks were fully charged. We made sure all bulkhead flappers were closed and all watertight doors between the compartments were shut and dogged (tightened) and blew ballast, which brought us to the surface.

While all this was going on, everything was calm with the officers and the crew. We all concentrated on the jobs we had been trained for to get the boat safely to the surface. If we had descended from our test depth of 412 feet to 450, I wouldn't be here to write this. To say the least, it was a very close call.

Our near tragedy happened in the deep water where the nuclear submarine *Thresher* SSN 593 made its final and fatal test dive in 8,500 feet of icy water in 1963.

JAMES P. CREEKMORE

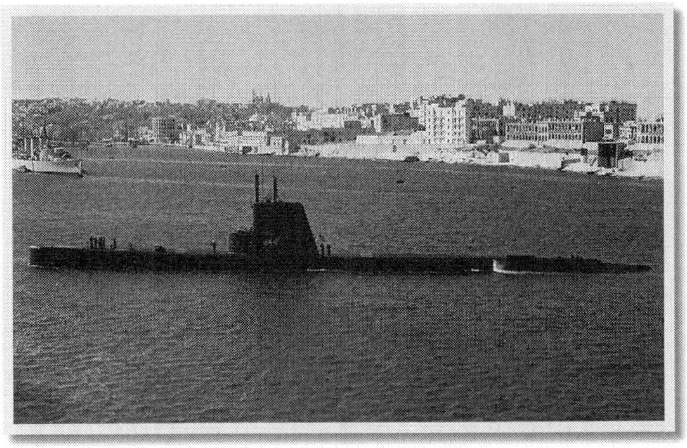

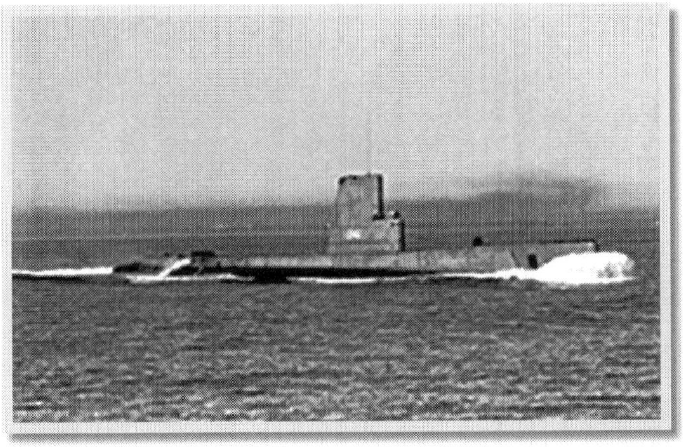

6 MEDITERRANEAN BOUND

On the morning of March the fifth, 1956, the Croaker got underway. We left New London headed to the Mediterranean. Our orders were to join Squadron Ten in the Sixth Fleet in a show of force near the Gaza Strip.

Getting to go ashore in the different countries before we got to Weymouth took away my thoughts of worrying about my car. As I think back, I think of the many more historic places I could have seen had I not spent my time with some of my shipmates chasing girls and drinking in the local bars.

We pulled in all lines that cold March morning and headed out from the State Pier in New London, down the Thames River, and into the broad Atlantic. The day before had been spent loading the boat with food and spare parts for the engines. There was no storeroom; the parts were stored in lockers and some of the big cans of coffee were stored in the engine rooms where they were fairly easy to get to and cushioned so they wouldn't make noise or tumble out on deck in rough weather.

Most of us thought that the first Port of Call would be in the Azores because of the problems we were having with our engines and our watermaker–the electric boiler that boiled our drinking water and the water used to fill the big batteries that let us operate below snorkel depth. The battery water had to be pure and be free of all salt; if the boiled saltwater wasn't pure, the battery cell being filled or topped off would explode, injuring or killing crewmembers and covering them with acid. But, we managed to solve the problem and continue on to the Rock of Gibraltar, and from there to ports on the coast of France, Italy, Greece, the Greek Island of Rhodes and the British Island of Malta.

The trip between New London and Gibraltar certainly wasn't a boring one for those of us working in the engine rooms. If we weren't standing watch, we were working on one of the engines. I, along with others who had just finished sub school, had plenty to do just trying to work on projects that we had to do to qualify. The Engineering Officer and Chief Engineman had no idea of the engine problems we would be having between New London and Gibraltar.

The engine trouble started before we got to the Azores when several cylinders in the After Engine Room started to overheat. The long run to Caracas, and now the long trip to the Med, started to take a toll on the engines, which had only been used on short school boat trips out of Groton and New London. The engines that had caused little concern before were now giving us problems. Most of the trip was made submerged at snorkel depth with the Engine Room gangs dealing with burnt valves and failing injectors.

During our greasy repairs, the most difficult part of the job was the removing and replacing the cylinder heads on the outboard banks. "Banks" was the terminology used because these were farthest from the center of the engine; the cylinder heads were close to the pressure hull. Replacing a cylinder head in the small space took more time because we weren't able to use the wrenches with the long torque bar handles; a special short-handled wrench was used to tighten the big nuts holding down the head. The tight space between the pressure hull and the engines left little space to work, and the torque-wrench–the wrench especially made for tightening down the outboard cylinder heads–had a very short handle with a loop at the end where a come-along was used to tighten it to the correct foot pounds.

I learned a lot on that trip, especially after working on the engines. I learned to repair injectors and worked on them on and off throughout the trip. By the time we passed the Azores, we had the engines pretty well repaired when another serious problem came up. This time it was the distiller. The Distiller Evaporator was used for distilling saltwater for drinking and showering and for keeping the 800-gallon battery water tank full.

When we started having problems, we quit taking showers and took sponge baths to save our drinking water. When we reached Gibraltar, we were a pretty motley crew and relieved when we hooked up to the fresh water on the pier and were able to shower.

That first day in port we went aboard H.M.S. *Eagle*, a British aircraft carrier, and loaded up with lots of soap, aftershave lotion and underarm deodorants. When we went to some of the bars like the Trocadero, we overheard some of the British sailors off of Her Majesty's *Eagle* and Australian sailors from other ships in the harbor talking and laughing about "the Yankee blokes" who were coming from the *Eagle* loaded down with deodorant and shaving lotion.

After spending several days on Gibraltar we got underway for other ports in the Med. We would get to spend more time on "The Rock" on our return trip from ports in France, Monaco, Italy, Greece, Rhodes, and Malta. We saw Crete through the periscope, but never went ashore. We did get some liberty time on Rhodes and Malta; by then we had been gone from New London for over three months.

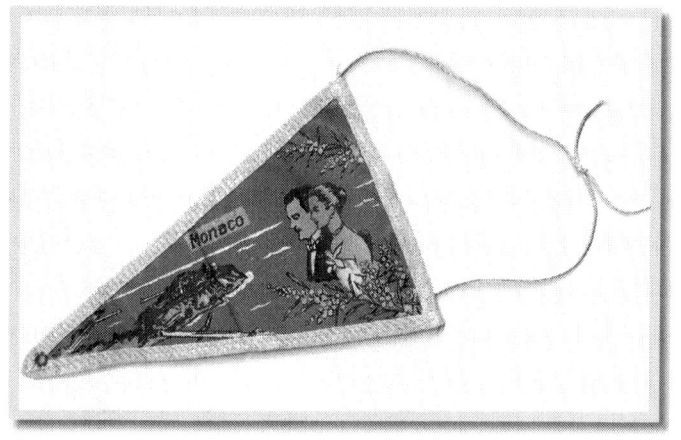

About six or seven miles from Rhodes, I got to see my first real serious communications using flashing light. Connolly QM2 (second class) was talking to the twenty-six or more different ships in the Sixth Fleet that were anchored off the island. I would later get to see the flashing light when we were having war exercises in the Caribbean. The bright light, which had shutters, was used to send messages in Morse code.

After leaving Greece, and three days after my twenty-second birthday on May the fifteenth, the *Croaker* softball team won the Sixth Fleet Submarine Force Softball Championship. The team had a barbecue and had plenty of beer before stepping up to the plate.

I, along with many others off the boat, didn't take part in the ball games, but still had fun while on liberty. We even had more fun when we went ashore in Valletta Malta.

Valletta was our last stop before heading for Gibraltar. The four days we spent in Valletta was another brief lesson in history, both in ancient history and the history of World War II.

Valletta had no vacant slips or berths we could use, so we spent the four days we were there tied to a buoy in the harbor. For transportation to go ashore, we rode the water taxi. The water taxi was a man-powered gondola like some used in the waterways in Italy, only this gondola didn't have the high bow and stern like the ones in Venice. We were told that the operator had leased the boat from a company in Sicily. It took a little longer to go ashore but the ride was worth it. Some of the drunken crew returning after their fling on the beach got a real chewing out by the officer of the deck as they boarded the boat. He saw them rocking the gondola as they teased the operator by swaying from side to side in the middle of the boat, almost causing it to capsize.

The merchants and townspeople knew that American sailors liked their beer cold and did their best to please us. All the street vendors and people wanting to make extra money and get a good price for their beer and ale had some kind of little ice chest at their feet while they held up their big brightly painted signs saying COLD BEER. I enjoyed my beer even if it wasn't ice cold. I especially liked Farson Blue Label Ale. It was at that time only brewed and bottled on Malta.

Valletta, like other ports we had visited, had a section of the city where there were plenty of prostitutes. One such popular section was what the British sailors called Gut Row. A bunch of us walked down the narrow alley and took pictures of some of the shops and of the prostitutes as they stuck their heads out the windows of the four-story buildings, advertising their trade to the sailors and civilians below.

On the third day, the day before we left, I had liberty and was topside waiting for the gondola when I noticed the British sailors aboard the English submarine *Sea Devil* that was tied to a buoy alongside of us. They were moving stuff around and taking some stuff below and coiling some of the lines and getting the submarine ready to get underway. I struck up a conversa-tion with one of them and he said they were getting underway, but only going about a half mile to tie up near two other English submarines, the Tip Toe and Tallyho, which he pointed about a half mile off our bow.

I couldn't believe it when he asked me if I'd like a ride instead of waiting for the water taxi, a gondola that was paddled or rowed. When I told him, "Sure, I'd like a ride," he said he would go below and get permission from the officer in charge. When he came back, he said I could make the trip and asked others standing on deck if they would like to go also. They said they had already had been aboard the sub and would wait for the taxi.

The World War I relic was like none I'd ever seen. I thought the *Croaker* had tight living quarters, but they were nothing compared to Her Majesty's *Sea Devil*. I was glad it was a calm day for my half-mile cruise otherwise my dress blues would have been a real mess from the oil in the engine rooms of the old sub.

Right off I noticed a box sitting in the control room filled with shot-sized, mini-bottles of rum. A crew-member said it was a daily shot of rum for each of the crew.

The ride was slow and smooth and was I glad when we reached the pier. It could have been messy, and even dangerous, in a choppy sea or rough weather I don't know what the test depth of the old submarine was. It couldn't have been over 300 feet. The bulk-head doors between the compartments were made of oak with stainless steel frames and the circuit boards or panels looked stable enough in their steel frames but they were open with no covers facing the deck, the walkway between the compartments.

If you were walking by one of them and the boat pitched or took a sudden roll and you slipped and lost your grip on the oak handrail, you stood a chance of sticking your hand in one of the circuit boards. You would get a serious shock and maybe be electrocuted, especially if your clothes were wet.

Walking through the engine rooms would also be a risk in rough weather and stormy seas. There were no valve covers covering the tops of the big Vickers diesel engines. I could plainly see why those that who had already been aboard decided to wait for the gondola. I now knew and understood why the crewmembers got their much-appreciated shot of rum.

I didn't ask, but I think the *Sea Devil* was going to be put out of service or decommissioned.

Return to Gibraltar

Our second stop at Gibraltar was every bit as much fun, if not more, than the first. For one thing we already knew where the gift shops and best bars were. We had more time to shop for souvenirs and again go sightseeing. Most of the non-rated sailors like me didn't have a lot of money to spend. I had fifty dollars less each month because I sent fifty dollars of my sixty-dollar-a-month hazardous duty pay home to Mother to help her make house payments and pay bills.

Main Street — Gibraltar

Before checking out the shops, I got a fast haircut by a Gibraltar barber. He did the whole haircut, not using an electric clipper, but a pair of some of the biggest shears I'd ever seen in any barbershop. I was quite relieved when I stepped down from the chair. I was glad I ended up with no nicks on my ears or neck. Af-ter leaving the barbershop, I met with some crew-members to check out some of the shops. I saw some intricately carved, hardwood figures that I wanted in one of the shops, but didn't have the money for them. I lucked out when one of my shipmates bought them and I later bought the carvings from him for fourteen dollars.

We were all excited when we left Gibraltar. We were in for some good times in the towns and cities on both sides of the English Channel. We were now going to see the Atlantic side and visit more places in France, see the White Cliffs of Dover in the English Channel and go on liberty in cities on the coast of England. We got to visit Weymouth and Bourne-mouth, which was the larger of the two and had a population of about 150,000.

The ancient Borough of Weymouth was the port where there was a gathering of submarines and support ships that had been on patrol near the Gaza Strip and those that had been patrolling the Atlantic side in October 1957. The ancient port of Weymouth in the year 1347 supplied King Edward III with ships and men, and in 1944, it played a very important part in the invasion of Europe by the allied armies. It was one of the principal embarkation ports. I had no idea at the time that Weymouth wasn't far from the Norwich area where the Creekmore heritage started in the year 1621.

A pamphlet was passed out to us before we went on the beach. Below is a list of support ships and submarines at Weymouth from that pamphlet: The support ships were the USS *Grand Canyon*, the USS *Shasta*, the USS *Papago*, and the USS *Fulton*. Sub Tender for Squadron 10 out of New London included 23 submarines: The USS *Sea Poacher, Bergall, Trigger, Jallao, Darter, Nautilus, Quillback, Halfbeak, Becuna, Trumpetfish, Sea Owl Chopper, Cavalla, Angler, Croaker, Grouper, Piper, Pompon, Ray, Redfin, Barbero, Runner, and Torsk*.

Weymouth

Several shipmates and I traveled from pub to pub in an old, four-door Dodge with wooden spoke wheels driven by a cab driver wearing a snap-brimmed cap.

He was very pleased to take us "Yankee blokes" to the pubs where we enjoyed the warm beer from the pumps. The pumps, as they were called, were actually air pumps that were used to put pressure on the beer in the wooden kegs behind the bar. When we traveled in the big city of Bournemouth, we rode in double-decked buses to the pubs and places of interest. Some of the pubs had rooms above the bar and dance floors where we could have even more entertainment—where we could enjoy the pretty girls.

The girl of our choice would lead us up the stairs to their room where we listened to old American recordings made in the 1930s and 1940s of the popular mu-sicians of those times, everything from Country West-ern music to American Swing. We listened to the rec-ords that were played on their old hand-cranked Victrola record players as we sipped their favorite drink called "Orange Squash," a drink made of orange juice and Scotch whiskey. We listened to the sounds of the low playing music as we slowly undressed for a fun time in bed.

When we left Weymouth, we crossed the English Channel to the Atlantic side of France. We were now stopping at seaports on the coast of Normandy where the bloody battles of Dunkirk took place in World War II.

Le Havre was our first port and the next was Dieppe where we would get to take the train to Paris and go on tour to the Palace of Versailles with its five hundred rooms and famous Hall of Mirrors and its huge acreage of gardens of flowers and fountains. It was quite a palace. It was originally built in the late 1600s and had survived through WWII.

Tapestries, big thick carpets that hung from the ceiling to the floor, separated the rooms. It had been left as it was originally built. Each room, depending on how many people that were living in the room, were equipped with chamber pots that were changed by the servants. Using long handled sticks with a hook on one end, the hooked the chamber pots and drew them between the thick, carpet-like tapestries that separated the rooms.

We also got to see the Eiffel Tower and take the elevator to the top, which was covered with clouds. I didn't have much spending money and didn't take the trip to the very top of the tower; I saved what little I had to spend in Pig Ale or Pig Alley, a part of Paris where we could drink and dance with pretty women.

While we were in Pig Ale, we stopped at one of the bars that had the longest copper bar top that any of us had ever seen. We had been served drinks and hadn't been sitting there ten minutes sipping our drinks when a fight broke out about halfway down the bar. Some of us at the bar had just bought watch fobs from the before the two got into a fight.

They screamed obscene words at each other in French as they got into it. They must have been known troublemakers because all at once two husky bouncers grabbed them from behind just as the two having the argument had pulled switchblade knives on each other. The knives fell to the floor as they were shoved through the swinging doors to the sidewalk and held until the police arrived. Never a dull moment for the *Croaker* mates.

Home Port

We were all happy when we returned from the Med and set foot on shore at the State Pier in New London. We were back to our homeport.

Married men were especially happy. Their wives and families were on the pier, anxiously waiting for the last lines to be secured, so they could be taken in the arms of their loved ones as the sailors stepped of the boat with their sea bags full of gifts and souvenirs. Some of the children were babies and toddlers when the boat left New London for the Med. We who were single, and sailors that had families living in different states, mailed our gifts home as soon as we could get to the post office.

After a month or more of operating out of New Lon-don doing patrols off the coast of Halifax, we received orders that our next patrol would be in the freezing waters north, maybe off the coast of Iceland or Greenland. After hearing that, we filled our lockers with foul weather gear. A couple of weeks later we were surprised and relieved when our orders were changed. Instead of going to the dreaded freezing waters of the north, we would soon be getting underway for the warm waters of the Caribbean, a part of the world that most of us were anxious to see.

We would be doing war exercises off Cuba, Bermuda, the Virgin Islands and Puerto Rico. We would be going ashore in Havana when it was still under control of the dictator Bautista who was running the country before Fidel Castro took over.

Liberty In Cuba

Before any of us went ashore, we all gathered topside and were briefed on what to expect while ashore. Our orders were to be careful and stay out of trouble and not give any excuse to be picked up by any of the green-suited policemen or soldiers who stood guard near the entrances and exits of every bar and gift shop, restaurant, laundromat, and whore house. The soldiers were everywhere.

When we arrived at Havana and cruised into the harbor, we found that most of the docks or birthing plac-es were already taken and we ended up tying up alongside of some American ships among the Cuban fishing boats tied to the piers in the brackish, foul-smelling water behind a big cannery.

The enlisted men going ashore used the same fresh water showers used by the grimy, smelly fisherman and cannery workers before we headed up the narrow alley to the street to catch a cab or bus for a tour in a better section of the city. Never in my travels (except later in the Mexican border town of Mexicali) had I seen such a raunchy, filthy display of human flesh as what was displayed in little rooms in the narrow alley leading up to town.

There was a line of little rooms, or I should say stalls, housing one prostitute per room. The rooms displayed half-naked prostitutes with their scabbed and pockmarked faces and legs, and were each furnished with a bed with a pull-curtain for the customer. At the foot of the bed, a dingy gray towel hung from the curtain rod where a bucket of water was placed for him to clean up in or to throw up in, whichever the case, before or after having his affair. This was like "Whores Row," all set up for the sex-starved fishermen who had spent weeks at sea loading the holds of their boats with fish for the cannery.

When we got to street, we had no trouble catching a cab that took us to nice, clean places where there were there were pretty girls, and where they served plenty of good, strong Cuba Libras (double shots of rum and Coca-Cola with a squeeze of lemon). We were getting to see things never seen before in any of the cities back home in the States or in the Med. The main part of Havana was quite different, and much cleaner.

There were some beautiful women in the nice, clean places we visited while we were there. Thank God, they weren't all like the whores we saw in the little rooms in the alley behind the cannery.

In one section in downtown Havana, there was a big shopping mall. The big, circular mall at ground level displayed a large fountain surrounded by benches where you could sit and relax and enjoy the cool breeze coming off the spouting water. As we made our way to the entrance to the different shops, the sidewalk started taking a gradual, corkscrew-like twist that wound its way up to the different floor levels of shops. As we walked uphill to the next level, we could see over the protective handrail the big fountain spurting cool water into the pool below it.

At each level, the shops lined the solid outer wall, displaying clothing, souvenirs and just about anything that we could afford to buy. Years later, this type of mall would become popular in the US.

About the second night of liberty, I had a little too much rum. I came back to the boat pretty drunk and stumbled and fell into a pile of scrap metal and skinned my right hand as I made my way toward the gangway of the ship I had to cross to get to the *Croaker*.

When I stepped off the gangway onto the ship, I lost my balance again just as I saluted the officer on deck. I fell against him and got blood on the shoulder of his white uniform. I remember him yelling down at the duty officer on the *Croaker* as I crossed the ship to go down the gangway.

"Come and get this man off my ship or he's going to end up in our brig below." The officer on the *Croaker* sent the topside watch up the gangway to steady me as I stepped aboard. I went below as soon as I got aboard and bandaged up my skinned hand and went to my bunk in the After Torpedo Room. Luck was with me, I didn't fight with anyone or damage any-thing other than getting blood on my uniform (and the uniform of the officer on the surface ship); so, I didn't get in trouble. The next day, as I was drinking black coffee and taking aspirin to treat my throbbing headache, I did get a good talking to by Chief Mann, the chief of the boat.

A couple of days later, after leaving Havana, we joined the Sixth Fleet, which was not far from Bermuda. This is where our serious war games began.

We were in constant contact with other submarines and all kinds of surface craft warships. At that time, the most sophisticated airplanes with sonar systems were the P2V Neptune planes. They were easy to recognize when you saw one fly over while standing on deck in the daylight. They had long, pointed tails that housed their sonar.

The clear waters were quite a change from the brackish water we had we had left when tied up behind the cannery. We spent most of our time wringing wet with sweat in a nice part of the Caribbean. We were constantly practicing drills; damage control drills, depth charge drills, you name it. There was a drill for any and all situations pertaining to any emergency or action while submerged or on the surface. We were continuously being called to battle stations. We were in an ideal area for the fleets to practice war games for the action we would take against the enemy.

For we who worked in the engine rooms, it was a tough job just putting up with the heat, especially when we rigged for "silent running" and the fans and air conditioning motors were slowed to minimum speeds. When approaching the convoys, they were turned off to keep from being picked up on their sonar.

We would come up to periscope depth, get our bearings on the ship (or ships) we were going to sink, and set the depths on the torpedoes with the TDC (Torpedo Data Computers) in both forward and after tor-pedo rooms so they would run well below the keels of ships in the convoy we were sinking in our mock attacks.

We fired a spread of three torpedoes at each ship we targeted to sink. The ships in the convoys would know that their ship had been hit when they saw the trails of bubbles from the spread being fired at them, the trails of bubbles trailing the torpedoes as they passed under their keels under their bows, amidships, and sterns. Within minutes after we scored hits on our targets, after firing torpedo spreads from both bow and stern tubes, the pinging of sonar bouncing off our hull could be heard throughout the boat as they zeroed in on our location. We made every maneuver possible to escape the onrush of destroyers and other depth-charge-carrying surface craft. With all of us at our battle stations, we went all out to run deep and silent.

The pinging of the sonar bounced off our pressure hull as plaster-loaded grenades were dropped by destroyers in the convoy. After a destroyer and a P2V Neptune got a fix on our location, we changed course and increased our depth. These torpedo attacks were carried out in the daylight hours and each torpedo that was fired by a submarine was retrieved at the end of its run and hoisted back aboard. Men went over the side and hooked up each torpedo that bobbed up. It was pulled out of the water with the portable hoist and lowered below where they were then rolled back in the cradles to be reset in the forward and after torpedo rooms. We were set up for the next attack. .

Liberty in Bermuda

When we tied up in Bermuda we soon found out that there wasn't much offered in the way of recreation for the single sailors who craved the company of pretty women, which we had enjoyed in other ports. There were plenty of pretty women all right, but most were with their husbands and friends from the big yachts like those that crowded the harbors in St. Thomas and St. Croix. It was Cinderella liberty for the sailors whose rate was below the rate of first class. Cinderella liberty meant that we had to be back aboard ship by midnight.

There wasn't a lot to do on the island except go to the nightclubs and bars or go scuba diving or swimming in the surf at the nice white sandy beaches. Or, rent one of the smoking Whizzer motorbikes and ride up and down the steep hills between the clubs and bars, which were a lot of fun.

Five of us went on my one and only short ride. Three were off the *Croaker* and the two others were surface craft sailors. Riding on the left side of the narrow, winding road was a little confusing after being used to driving on the right side back in the States. Even more so after a few beers.

The surface craft sailor was the first one to pull out of the rental lot onto the narrow road and he almost collided with a car in the right lane. He took off ahead of us leaving a trail of smoke, going full-throttle before pulling off at a turnout on the left side, about a half-mile from where we started. We just got a glimpse of him and then, all at once he was out of sight. Now you see him, now you don't.

When we reached the turnout where he had stopped, there were a dozen people in a group watching as oth-ers were pulling the bike up and helping the sailor climb back up to the road. In his drunkenness, he didn't realize he had coasted too far over on the shoulder after he turned off the engine. When he put his foot down, his leather shoe sole slipped on the slick grass, causing him to lose his balance and both he and the motorbike slid about thirty feet down the slope on the slick grass.

He was lucky he wasn't seriously hurt other than skinned elbows and knees. He was lucky too that the motorbike wasn't damaged. One of the sailors in the crowd rode his bike back to the rental place and one of the other surface craft sailors who had started out with us saw to it that he got back to his ship.

After that sobering experience, we only made a couple of stops before returning the rented motorbikes. One place we stopped was a bar or nightclub, and it was a place that didn't greet us with a hardy welcome.

When we parked our motorbikes in front, we could hear piano music and laughter as we made our way inside. As soon as we stepped inside, everything got quiet. It didn't take us long to find out why. After our eyes adjusted to the dim lights, we saw the hateful looks on the faces of people sitting at the tables, some had turned around on their stools at the bar and were giving us hateful looks. I guess we really stood out when we walked into the smoke-filled room. We were the only white people in there. It didn't take much coaching from the bouncer who came over and told us politely that we best leave.

We walked out and were very glad that none of us had said anything to get us in trouble. We were told later that some of the islanders had their own recreational things and didn't mix with the tourists.

Several of us went out on a charter boat and went snorkel diving. We couldn't afford to rent gear for a scuba dive. The snorkel gear rental was much cheaper and we didn't have to swim with a supervisor to make sure the regulators on the tanks were properly set.

Before getting in the water, we were warned of the some of the dangers, some fish and shells and coral to stay clear of and not walk on. Our gear was swim fins, a snorkel mask, and a net bag to put our shells in as we gathered them. There were some beautiful reefs with all kinds of marine growth with different varieties of fish. The skipper of the boat knew just where to take us where we could easily see the bottom to make our dives to gather our shells.

Everything seemed to be magnified when looking through our snorkel masks in the crystal clear water. Before starting back to the charter boat, me diver and I swam out to the deeper water at the edge of the reef just to see how deep it was. We decided to go no further when we saw a big ray resting on the bottom. The clear water's magnifying effect made it look much bigger than it probably was. We weren't about to go down and check it out.

Before the trip was over, all of us managed get some nice shells. I lucked out and came up with one of the biggest shells–a nice, big, pink conch shell. I cleaned it up after flushing out the sand and tiny crabs.

I got to go on liberty quite a bit while we were there. One afternoon, Gene Kirkland (Kirk) and I were invited to bring his guitar and my mandolin to a bar and have a jam session with the piano player. We had been in the bar a couple of times before and the piano play-er, a black man who had been raised on the island knew a lot of songs that Kirk and I played. The bar, which was within walking distance, was filled with sailors off the *Croaker*. Bill Hickok (Wild Bill Hickok), a chief engineman, was among the crowd. He knew the words to most of the songs we were playing and began to sing with us as we played. We were having a lot fun playing music, drinking, and telling jokes and we hadn't been watching the time. It was eleven-thirty when we walked out the front door. We were just leaving as two shore patrol walked in to make sure all under the rate of first class left and headed back to their ships. We had to get going; the walk back was long.

The shore patrol had just left when Hickok came out the back door with two sailors carrying cases of beer and yelled at us to wait up. We started walking the trail we had taken when we carried our instruments down to the bar. The trail went around a hillside that was covered with short tree stumps, spots where we sat to enjoy our beer. Kirk and I found a big stump, broke out the mandolin and guitar and started to play when Hickok told us that he wanted us all to sing together. He said he used sing in what was known as a barbershop quartet. It didn't take him long to line up the group to sing in harmony.

The last song we sang before we went back aboard the *Croaker* was an old tune called "Heart of My Hearts." Here are some of the words, the few I remember.

> *Heart of my hearts, I love that melody.*
> *Heart of my hearts, brings back old memories.*
> *When we were kids, on the corner of the street.*
> *We were rough and ready guys, but oh how we could harmonize.*

> *Heart of my hearts, brings back those memories.*
> *I know the tears would glisten, if only I could listen*
> *to that gang that sang, heart of my hearts.*

After a few tries, everybody did a good job singing it. It was a tune the drunken sailors and I will never forget. I can't remember all the words; but I'll never forget the tune. I'm sure the sailors that stood on the hill singing that night will never forget it either.

War Games

During the war games, many others and I got to see things we had only seen in the WWII movies in boot camp. We again got to witness what we had seen when we were with the Sixth Fleet off the Island of Rhodes, the signalmen sending code signals by flashing light to the various ships and submarines in the fleet, some offshore and some in the harbor–Morse code sent by flashing light.

Two men stood on the bridge near the compass. One was using the underwater phone to communicate with ships that could not be contacted by the light. You would see the signalman give three or four flashes, rapidly opening and closing the shutters on the light, then pause a few seconds and repeat. I guess the thrill we got out of this was standing there in total darkness looking out toward the fleet and watching the flashing lights, the returning coded messages coming from the ships the signalman had contacted in different positions all around the harbor.

Traveling the Caribbean was an experience all its own. I just wish I could have spent more time topside instead of spending most of my time in the sweltering engine room. What time I did get to spend up on deck in the fresh air wasn't nearly enough. I missed out on taking pictures of the small ferryboats as they plowed the waters between the islands. The small, privately owned ferries were sixty to eighty feet in length and powered by both sail and diesel engines. The boats were haphazardly loaded with all kinds of cargo. Covering their decks were a mixed bunch of anything and everything. There were cars and trucks and small pens or stalls crowded with baby goats and large cages with different varieties of parrots and other bird species gathered from different islands.

One such ferry passed us while we were standing on deck in the afternoon. Of all times, none of us had a camera to take pictures of it. As the old saying goes, one picture is worth a thousand words. As the boat passed not thirty feet off our starboard bow, it was heeling over and listing so bad with its offset, unbalanced cargo, we thought for sure it would capsize before we could get past it. We were almost sure we would be trying to maneuver our way between a flotsam of drowning, bleating animals and cages of squawking birds.

After our short liberties and fun times in Cuba and Bermuda, our next island stop or "port of call" would be San Juan, Puerto Rico. It would be our last island to have fun on before heading back toward New London. San Juan was a fine place to enjoy plenty of rum and have some real fun in the company of more pretty women before heading back to the States.

Like the other islands in the Caribbean, San Juan had beautiful white sandy beaches and tall palm trees. I only went on liberty three times while we were there. The second time, four of us took a cab ride to a place on the island that was noted for its "clean girls." A place the cab driver said was one of the cleanest on the island.

It was late in the day when we finished having our fun with the pretty women. The sun was low in the sky when we left through a side door and headed for the cab waiting in the parking lot.

We heard voices speaking in rapid Spanish that became shouts as we made our way toward the cab. We stopped about twenty feet away, as the arguing grew louder and louder. We saw that there were two men arguing with the cab driver.

We realized why the driver of the cab had started talking in an apologetic tone when we saw what looked like a carpet cutter, a knife with a hook-like blade, in the hand of one of the men he was arguing with who was standing close to the cab. We knew what the argument was about when the driver pointed to the four of us.

The Puerto Rican with the knife wanted the waiting cab to take him and his friends to back to San Juan or some other part of the island. The argument was over when I yelled at the driver and told him we would call another cab. The man put the knife back in his pocket, and two more men joined them as they piled in the cab and drove away.

When we got underway from San Juan that March in 1958, I only had a few months left until it was time to re-enlist or be discharged. I found out I had to do an extra month to make up for the time I spent in the brig.

Our trip home, back up the New England coast, turned out to be a slow and very rough ride as our big bow plowed through the pounding waves on the sur-face of the stormy sea. We were tired of the war games and were looking forward to being back in New London where we single guys could be with our girlfriends and the married guys could be with their wives and families. I didn't know at that time that Nig Ross, Moldy Higgins, and Ralph White, three sailors that had served on the *Croaker* in WW II, were also getting discharged. Nig Ross was always teasing Higgins, talking to him in a feminine voice, calling him by his first name, which was Francis.

7 DISCHARGED

The day I was discharged I had planned to go home for a while, go into the active reserve, and then go back into the Navy and go to nuclear power school and finish my career on a nuclear submarine. The last time I had been home was when the Croaker was in the Navy yard in Philadelphia having her batteries replaced.

On my way home, I planned to stop in Baltimore, see my brother Bill and his family, and travel on from there to Wewahitchka, Florida to see Lila, Phil and their handsome little son Bill, and from there on to Victoria, Texas to visit Laura and her family, her husband Johnny and their two pretty little children, Michael Patrick and Laura Jean. From there I would head home.

In 1956, the family–Mother, Dad and Tom–had moved from Burns to San Jacinto, a little town near Hemet in southern California. So, when I left Texas, I wasn't heading to what I called "home." It seemed like my best plans never worked out. Here I was getting off the bus in a little out-of-the-way, dried-up town in the desert.

It turned out, though, that this time things did work out for me. In the little town of San Jacinto, I would meet Jean Standard, a beautiful little woman who was all of 4 foot 10 inches tall, who two years later would become my wife. It was the best luck of my life.

The Greyhound bus ride from Victoria, Texas to San Jacinto wasn't much fun. I had a window seat and there certainly wasn't much to see while traveling across west Texas, southern Arizona and southern California. I could see why the Texas cattle ranchers having large herds of cattle had to have such big acre-ages. There were miles and miles of rocks and flat-lands with scrubby grass and weeds. Having land like that, you could plainly see that it would probably take ten acres to feed three or four cows.

I read magazines and slept the biggest part of the way. I did have some fun taking time off to go into Mexico when the bus stopped in El Paso. We got to El Paso before noon and I decided to go to Juarez, which was right across the bridge over the Rio Grande River. I would have a little fun and be able to catch the eleven o'clock bus out the next night. I was traveling in uniform and had no trouble visiting and dancing with the girls in the bars. The only other time I had been there was when we were living in Canutillo. In 1947 Laura and Leta took us to El Paso to see for the first time live alligators that were downtown on display in a big circular cement pit that included a big pond and plants around it like their natural habitat. After seeing the alligators and looking in some of the shops, we got to enjoy a nice Mexican dinner just across the bridge in Juarez.

The last Greyhound stop was in Riverside. From there I rode the Blue Mountain Stage, a small bus that ran from Riverside to Hemet. The bus traveled west on Interstate 10 toward Palm Springs and then cut off on Highway 79. Highway 79 was a two-lane road that went southwest to the little town of San Jacinto and on to Hemet, about thirty eight miles from Riverside.

When I stepped off of the bus in downtown San Jacinto it was blistering hot. I shook my head in disbelief; I couldn't believe these folks had chosen such a dried-up, out-of-the-way place to buy a home. Why in the world had they moved to San Jacinto?

I called home and stood in front of the Tack Room, a small bar, and waited for Dad to come and pick me up. I watched the senior citizens driving up and down the street in their electric cars, cars that were nothing but golf carts with canopies to keep the riders from frying in the broiling sun and tall poles on the backs of them flying flags of caution so they would not be run over by the normal traffic.

Mother had called Tom at the service station on State Street where he was working and they were both at the door to greet me when we pulled into the driveway at the back of the little house. It was sure good to be home.

I found out later that Aunt Minnie and Uncle Grant and family lived in Bell Gardens in East Los Angeles, and that Mother's half-sisters, Mildred and Willie, and their mother, Ma Jones, lived in Pasadena. Will Jones's second wife who we called Ma Jones was very sick. She had come out from Kentucky and was living with Mildred and her husband John Wood. Mother and Aunt Minnie had visited them and mother said

Ma wasn't expected to live and could pass away at any time. She died a short time later.

That fall Bill and his children came out from Victoria and were going to rent a cabin, but Mother wouldn't hear of it, not just the three of them without their mother. Dad was working nights drilling box upon box of small metal parts in his little machine shop, some the size of dimes that were drilled for the elec-tronics plant out of Pasadena where Mildred's hus-band John was manager.

A glass-like ceramic was baked around the holes in the parts before wires were thread through them for the electric circuitry. Dad worked from late afternoon until dawn, and slept in a bed in the small washroom that was connected to his shop. Bill worked with Dad helping him fill the orders until he got a job tending and managing the bar in Mountain Center on Mt. San Jacinto above Palm Springs.

He worked there several months helping support the family before taking a job in L.A. At the time I was working in Banning between San Bernardino and Palm Springs in a body and fender shop with Earl Mullens, a guitar player in Tom's band. I only worked there a month.

One weekend, cousin Dave had some time off from his job and came out from Bell Gardens for a visit. Bill was bartending part-time in San Jacinto and had didn't have to work on Sunday or Monday. Dave called Bill and said he would help pay for the gas if Bill, Tom, and I wanted to spend the weekend in Mexicali. He said there would be more room for the four of us if we took Tom's four-door Mercury.

After Bill got off work about two-thirty Sunday morning, we piled in the car and headed for Mexicali.

About forty miles from Mexicali, we were cruising along about seventy when all at once we heard a big bang, like something had hit the floorboard, and the engine began to knock. Tom pulled off on the shoulder.

When we got out and looked, there was a trail of oil going down the narrow highway to where the rod piston bearing went out, driving the rod through the pan, and sending it flying out into the field.

Needless to say, we all felt sorry for Tom. He was really disgusted as we pushed the car into a farmyard. The engine was shot. The Mexican man that came out of the house was very polite as Dave, who spoke fluent Spanish, explained to him what had happened. The farm worker said it would be okay to leave the car there until we could have it towed home.

We had started walking toward the road when a guy in a Volkswagen who had seen from the oil trail what had happened, stopped and offered us a ride if we could all manage to squeeze into the little Volkswagen. We all managed to get in. Dave, being the smallest, sat partly on Tom's lap.

We spent Saturday night touring the bars before pooling our money for bus tickets to get us back home Sunday. Dave gave each of us a peso when we got back to remind us of our trip to Mexicali. (I drilled a hole in mine and had it on my key chain for a while before I put it up as a keepsake.) The following Tues-day we helped Tom pay the towing bill to get his car home. Not long after that, I helped him replace the engine with one he and I had rebuilt.

I became restless after being home a couple of weeks with very little to do but fix things around the house. I had visited the local bars trying to get acquainted with some of the locals, asking them about some place close where I might apply for work. I had very little spending money. The $700 ("mustering out" pay) I had received before I left New London was pretty much spent and I had to find some kind of local job.

Not far from the house was Jim Minor's potato shed. The sound of the conveyor belts creaking and the drone of the trucks pulling on line to off-load the potatoes, the flatbed trucks loading tons upon tons of sacked potatoes after they were culled and sorted and sacked according to size, made constant noise that could be heard throughout the night.

One sleepless night at about two o'clock in the morning, Tom was snoring away, so I got out of bed and went into the living room to try to get away from all the noise. The small bedroom that Tom and I shared with two small beds and a chest of drawers was facing the street and the noise of the activity in and around the potato shed was coming in through the open window. We turned off the swamp cooler and kept the window open during the night to keep from running up the water and electric bills.

Tom was used to the noise down the street; but being the light sleeper I was, it kept me awake. I went back in the bedroom, got dressed and went out and sat down on the front porch for a few minutes before I decided to walk down and check out what was going on where all the noise was coming from. As I ap-proached, I could hear someone giving orders in both Spanish and English. I had been told by the people uptown that they did hire people at different times, but the wages were pretty low except for those sacking the potatoes and sewing the sacks. I was told too that you really had to work hard and fast to keep up, and that those jobs were hard to come by.

When there was a little break I had a chance to talk to one of the foremen. He said to show up the following Monday, they were going to hire a cleanup man and he would put me to work cleaning up around the conveyor belts. The pay would be seventy-five cents an hour with no raise in pay when I had to work overtime (which I did every night).

They worked mostly illegals, Filipinos as well as illegal Mexicans.

By the end of my second week, I was really mad. The Filipino that I had been working with said he played the mandolin, but didn't own one. He had invited me over to his cabin to meet his family. At the time, I had two mandolins, a 1915 Gibson I bought in a pawn-shop in San Francisco on my return trip from Japan, and the other was a Silvertone that I had bought in Salem, Oregon when Dave Carroll and I worked in the bean fields in Silverton.

As my wife told me many times after we were married, I was always helping people who showed no appreciation. She said that I was too good to people, people that I really didn't know. She used the phrase, "You're too good for your own good." How right she was. I had loaned the Silvertone to the Filipino, not knowing he was leaving for parts unknown. Not even the foreman or the manager knew he was leaving. I found out when I went to his cabin to pick up my mandolin, finding the cabin vacant. He had moved out without telling a soul.

♪

Tom and I had planned on pooling our money and enrolling in Indian Valley Junior College in San Bernardino the following fall. He was going to major in music and I had planned on taking a course in diesel electric technology since I had already some experience in working on large diesel engines while in the Navy. When we did enroll that fall, we only went for two weeks. We found that after buying our books, slacks and dress shirts (the required dress code for students) was just too costly to make round trips to the college and back five days a week.

I had been working at the potato shed two weeks when Dave Carroll called me from Bell Gardens. He was working nights for C. E. Howard, a company in Downey that made insulated stainless steel milk holding tanks for dairies and stainless tanks for milk tankers transporting milk to the creameries. He had talked to his boss and was able to get me working on the same shift for a dollar an hour. He and his brother Grant both worked there.

My job was grinding down and smoothing off the rough spots on the welds inside the tanks and insulating them.

When we left for work, Dave and I started our day by having to push-start Dave's little Henry J Kaiser, which had a broken starter. It was dangerous even pushing the car out of the driveway, and even more dangerous when we got it out on the side of the busy street. The traffic swung out around us as we started down the street with Dave holding the door open with one hand, the other on the steering wheel, running down the street getting ready to jump in and pop it in gear when we got it up to speed. I would then run as hard as I could to catch up and jump in as the cars whizzed by.

I never liked working nights especially the ones at C. E. Howard. It wasn't the job so much as living with my cousins. I was treated with kindness as I always had been in my younger years. I had talked of renting a room someplace near the job, but Aunt and Uncle wouldn't hear of it. They insisted on me staying with them. It was much different living with the Carroll family than it was when I was a little guy. The kids were all practically grown and Uncle Grant was still as overbearing as ever with Aunt Minnie and his grown-up kids.

In their younger years, when he corrected his sons he whipped them with his razor strap. He was never physically abusive with Aunt Minnie, but was constantly yelling, shouting orders at her like, "Get me a cup of coffee, crow bait," calling her degrading names in front of all of us. I don't know how she stood it all those years.

The kids, the youngest ones in their teens, were in and out of jail for being on drugs and other things like purse snatching, a crime that cousin Donnie was convicted of. He was sentenced to work ninety days in Road Camp, gathering and bagging trash that people had strewn alongside the highways, and he had a second job fighting grass fires.

Dave and Nancy were the only two that lived decent lives. The three boys that were always in trouble were Grant junior, Donnie, and George, the youngest of the family. They stayed away from home as much as possible doing odd jobs, but hung out with the wrong people after work. After being there for a week I could fully understand why they didn't want to be there. I really wasn't making enough to be able to afford renting a room in a motel and still be able to save something. I really wanted to leave Bell Gardens.

I gave Aunt Minnie money to pay for my food, but I didn't pay any rent. Grant was buying drugs and alco-hol with most of the money he made. Most of his drugs came from his friends that he drank with at the Chit Chat. The Chit Chat was an Okie bar that we went to for entertainment sometimes after work and on weekends. The bar had live music and plenty of girls. Ray McCoy, a guitar player who really drew the crowds, was like a grasshopper; one minute he would be walking up to the tables playing his guitar, the next he would be up on the bar singing the latest country western tunes while the rest of the band was onstage with their amplifiers going full blast, playing in the background. How he ever kept from getting tangled in the cord that ran from the amplifier to his guitar and breaking his neck, I'll never know. One of the popular songs he often played and sang was "Pick Me Up On Your Way Down."

The trashy lifestyle and coming in contact with some of the drug addicts that Grant, Donnie, and George hung out with was all the more reason for me want to leave Bell Gardens. I couldn't wait to get a couple paychecks and get out of there.

One Monday morning, I took a bus ride to San Pedro. I had talked to a Merchant Marine sailor in the Chit Chat and was told that with my Navy experience I might be able to get a job in the boiler room on one of the oil tankers. He said I would make quite a bit in overtime money by standing watch for those who wanted to spend extra time on the beach as they waited for the big tankers to have their tanks filled with crude oil.

I picked up all the necessary papers and a small book to study up on survival at sea and the other things that

I would need to know when I took the test and got my ID photo. My name then went on a waiting list.

I thought if I made a few trips to the gulf, I would be able to save enough money to get a nice car and drive back to the east coast if I decided to reenlist and go to nuclear power school. My other plan was to save enough money for Tom and I to go to the junior college in San Bernardino. I went home a week later and had a complete change of plans.

Dave was busy working on his car the Saturday Grant and I went home to San Jacinto for the weekend. I did the driving. I didn't trust his driving with the pupils of his eyes looking like the eyes of a hoot owl, dilated from the effect of the Benzedrine pills he had taken the night before. I knew he wouldn't fall asleep, but I wasn't about to let him drive me home. When we got home, Mother noticed he was on something right away when she met us at the door.

She called me aside and said she never wanted to see him in her house like that again. She said she wished that he would get back in his car and go home. We really had no idea then just how many different drugs he may have been taking besides the Benzedrine pills.

That Saturday evening, I got him out of the house to get him away from Mother. I had him walk with me to the main part of town in San Jacinto. I was going to try to find something to make him sleep so we could send him on his way back home on Sunday. We walked uptown and stopped at the first bar we came to, which was the Ramada Inn. I opened the door and we stepped inside the noisy bar. The Ramada Inn had live music on Saturday nights.

I left Grant sitting at the bar while I walked over to a pretty little woman sitting at a table near the band members who were picking up their musical instruments as they got up on stage, getting ready to play.

Jeanie

She was beautiful. She stood out, obviously the prettiest one sitting at the table.

She sat there talking to several women who were seated at the table near the side of the stage. The band had just started playing as I walked over to her and asked her to dance. She was perfectly dressed in a perfectly starched and pressed, high-collared, snow-white blouse covered by a little starched and pressed blue vest, Torador pants and black shoes—everything perfectly matched.

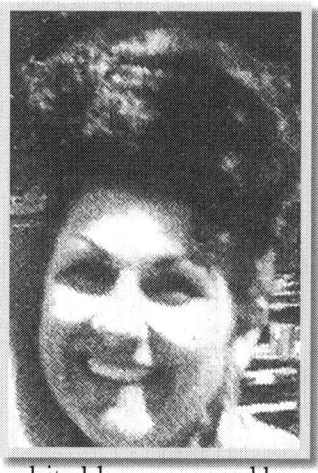

I found out after the first dance that she and a couple of the other women at the table were dating the members of the band. She told me that she had come to the Ramada with Chuck Cline who was the lead musician. She said that she and Chuck had just started dating. Chuck was a very talented guitar and banjo player. I later found out Chuck was a religious guy who spent a big part of his time playing gospel music at church and for a time had performed with the famous Blue Grass singer and mandolin player Bill Monroe before traveling west to Hemet.

Chuck and his band had no trouble playing the latest popular songs. I was standing in front of the stage holding Jeanie's hand as Chuck and his band walked on. After they all took their proper positions, Chuck looked at me and asked me if I had a request. I asked him if he would play "Satisfied Mind."

Once we started dancing, we danced to every song and took a break only when the band did. We got pretty well acquainted as we danced every dance until closing time. The tiny little black-haired woman all of four foot ten inches tall was as pretty as a doll and an excellent dancer. I knew right then I would definitely have to see her again.

The next day I told Grant I might quit C.E. Howard and move back to San Jacinto. Before Grant left the house that Sunday, Mother made it quite clear to him that she didn't want him coming to the house again until he straightened himself out.

I told Tom about the situation when he came home from work that afternoon and after talking to him, I knew I was moving back for sure. He told me that American Pipe was hiring laborers and I stood a good chance of being hired. They were making sections of big concrete pipes for the aqueduct. He said too that I had a chance of getting a union job working construction on the section of freeway they had started to build between San Bernardino and Banning.

I joined the Laborers Union and went to work for a paving company at March Air Base, my first union job. I had to be on the job at four in the morning. My job was helping fine grade the sand so it could be compacted between the metal forms or tracks in front of the big paving machine, and after a section had been poured the other part of the job was helping sev-eral laborers cover the big slabs with rolls of burlap that was then watered down to keep it from cracking while it cured out in the blistering sun. I made a dollar and twenty-five cents an hour and was paid time and a half for overtime. I worked from four in the morning until twelve and sometimes one in the afternoon.

Most of the laborers that I worked with had at least completed grade school or the eighth grade.

There was a black man about fifty years old wearing a silver hard hat with the name "General" painted on it. We soon found out that the poor man couldn't read a measuring tape, and as we all marked our shovel handles to help us grade while we spread out the sand between the forms, the foreman had to mark his for him.

It didn't take us long to see that the General's mind was warped. He was a strange guy, and we all begin to wonder what was going on with the man when we noticed that the trunk of his car where he kept his lunch was locked with a padlock, and when he pulled the lunchbox from the trunk, it also had a padlock. We found out later from one of the workers he rode to work with that he was quite paranoid. The General said he was sure his wife was trying to poison him and he made sure everything was locked.

One morning he couldn't get his car started, so he had his friend who lived not far away give him a ride to work. His friend, and one of the other workers that lived in Perris, said the General lived on five acres and his property had several junk cars alongside the dirt road of the driveway leading to his house.

As he was rounding the turn leading up to the house, he said he saw three small children duck down between two junk cars as the headlights came round the turn. We guessed that the family was scared to be around him and hid out until he left for work. He was a pretty nutty looking guy and no one, at least while I was working, asked him about his problems.

The job didn't last long because after I went to work there. The paving company was finishing up their contract for a section of airstrip three feet thick and eight hundred or more feet long. It was an extension of the landing strip for the big B-52 bombers. The bombers were taking off and landing all day long.

One of those early morning landings really gave us a thrill, or better put, scared the hell out of us. It was just breaking daylight when it happened.

I had just left the parking lot to join the rest of the crew who were walking down toward the big paving machine when all at once the ear-splitting noise of a big jet coming in for a landing startled us with its flashing lights and landing lights on, and the big jet engines revving up to break its speed. It was a sight to see, a most scary sight to see.

The big plane was coming in at such an angle we thought it would crash and explode. To our amazement and good luck it didn't. When the plane touched down, sparks flew as it bounced when the big tires hit the pavement skidding and screaming, filling the air with green smoke from the burning rubber, the huge wings flexing with their tips almost touching the runway as the pilot got it under control.

By then we were all hunkered down behind the paver. We had seen the big planes land, but never at that steep of an angle, and never that close. We figured he must have gotten the wrong landing instructions. That part of the runway was paved a couple of months earlier but wasn't being used.

When that job was over, Tom and I both went to work for American Pipe. By then, Jeanie and I were well acquainted and were going steady.

♪

Mother liked her the first time they met, which was the evening after Thanksgiving 1959 when I brought her by the house. We sat down at the table in the little dining room and enjoyed the leftover turkey and all the fixin's while Mother got acquainted with her.

Jeanie, being the fine cook that she was, just loved Mother's special dressing. And having been born in Arkansas and raised in poverty herself, Jeanie had been through the rough times in the thirties and forties, much like our family. Mother took to her right off.

When we got married, I still didn't have a nice car. I had to buy a car for work, so I bought an ugly old gray '41 Dodge coupe from the wrecking yard for $75. The person that had it before me had taken the trunk lid off and bolted together a plank box that extended out of the trunk and over the back bumper, making it into a pickup.

When Jeanie and I were first dating, we used her 1950 Ford, a nice two-tone. It was a brown and bronze two-door sedan with a nice sun visor. The sharp little car had a major engine problem, and a month before we were married I replaced the engine.

While we were going together it seemed like I was always getting into a fight over something or another, like the time I got in a fight with two guys in the Elbow Room in San Jacinto. One fight started when a guy who was sitting next to Jeanie found out that she and I were going steady. This made him mad and we got in a fight. I was lucky the bartender stopped the fight and we left before the police were called.

Here's how it started: Jeanie and I were having a few drinks in the Elbow Room when two men walked in and sat down at the bar. One of the men recognized Jeanie right off. He had tried to date her months before and she had turned him down. She told me after the fight that she wanted nothing to do with him. She said he was a loud-mouthed braggart. When the two walked in, they sat down next to her. It was early in the evening and there were very few people in the bar.

Jeanie sat there trying to ignore him as he was trying to get her attention, talking loud with the jukebox playing in the background. Jeanie had turned her back on him when he and his friend finished their drinks and were getting ready to leave when the loud-mouthed braggart turned around and yelled loud enough for everybody to hear:

"I don't know what you are doing dating a stupid son-of-a-bitch like him," meaning me.

I was furious.

I stepped down just as they turned to head for the door and punched them both in the face, knocking them to the floor, and I would have kept on beating them had the big bartender not come over and stopped the fight.

They both had blood running down their faces and the knuckles on both my hands were skinned and bleeding when Jeanie and I left the bar.

I was lucky. Dick Clevenger, the big bartender, was a good friend of Jeanie's and he didn't call the Hemet police until he was sure we were safely away.

American Pipe was the first and only job I was ever fired from. We had been working three weeks and the Sunday evening of our last week, I borrowed Tom's '51 Mercury, a nice cream-colored four-door.

Tom played sax and vibes at various clubs in and around Palm Springs on Friday and Saturday nights and sometimes when he rode with other members of the band, he let me use his car.

My first stop that Sunday evening was the Ramada Inn. I had planned to stop in for a couple of drinks and then call Jeanie to see if she wanted to go out.

I was sitting there having a rum and coke when I struck up a conversation with the guy next to me. He asked me if I could give him a ride to the bar in Sunny Mead, California near March Air Base. He said from there he would call his girlfriend and she would bring money to pay me for the gas I used so I could fill Tom's car with gas when I got back.

It was after seven when we left the Ramada after he bought us each a shot of 151 proof rum. We got to his destination all right, but were both loaded. I sat there having a drink as he made the phone call to his girlfriend and walked out front to meet her.

That was the last time I ever saw him. I was conned for a free ride. The bartender noticed how loaded I was and refused to serve me another drink. When I stood up from the stool I was sitting on next to the swinging doors going into the kitchen, I bumped a waitress getting ready to go in to fill an order and was immediately grabbed by two guys that literally threw me out in the gravel in front of the place.

I stumbled to my feet, and as I was trying to get the key in the ignition of Tom's car, I was yanked out of the car and stuffed in the back of the Riverside county sheriff's car and locked in, and while they were inside talking to the bartender, I kicked open the back door of the police car and was walking toward Tom's car again when I was then put in handcuffs, crammed back in the police car and taken to Riverside county jail.

At the time of my arrest I was wearing Bermuda shorts, Mexican sandals and a short-sleeved shirt. I was a drunken, skinned-up mess as I stood there in the drunk tank with a slip of paper sticking out of my shirt pocket with the numbers referring to the charges against me. By then, I had pretty much sobered up. I stood there with my head throbbing and watched the grubby bunch that were walking around in the vomit and practically fighting over the battered tin cups of powdered milk and bologna sandwiches that were brought in and placed on the metal table in the center of the room. That afternoon I was arraigned and charged for resisting arrest and damaging a police car. For both charges, I paid a fine of $700. There went my savings.

I was allowed to make one phone call while there, so I called home for somebody to come and bring me the bailout money. I considered myself lucky, I still had my driver's license and was able to drive Tom's' car home without a scratch.

Tom and I both lost our jobs with American Pipe because neither one of us had called the foreman ahead of time to let him know that we couldn't come in to work. We weren't too disappointed; the jobs he and I had were miserable.

The aqueduct pipe was poured in ten-foot sections. The big metal forms or sleeves with re-bar (reinforcing steel) cages inside of them were ten foot high, nine foot wide, and the walls of the pipe sections were a foot thick. The forms were also steam-heated so the cement would cure faster and that made the job all the more miserable for all of us. Tom and I would have to hurry up and down the ladder as each section was poured. Our job was to keep the portable, round platform clean so that when the pour was finished, we all could clear the platform and move on to the next section. We stood back out of the way while the cement finisher swung the big funnel-shaped bucket, which was lowered by the crane, and filled the form all the way around while another smoothed it down with a trowel. Tom and I took our square-point shovels and quickly cleaned off the excess cement around the form and threw it over the side.

My next job was doing pick and shovel work for Matich Brothers Construction. I had only been work-ing a week when I told the foreman that I had spotted for heavy equipment, a job I'd had before going into the service. I soon got the job as spotter or dump man.

My job working on the Interstate, or I-10, was much bigger than any construction job I'd had before. Most all of the equipment was much bigger and the operation much faster. I worked close to the big rubber-tired Hough payloader that was spreading the material as it was dumped. The tires on the big payloader were about as tall as I was, so I really had to be alert to stay in the clear as the operator went quickly back and forth pushing the fill dirt.

When we were working near San Bernardino, I spotted for forty-two bottom dumps, and four scrapers or big earth movers, and managed to stay out of the way of the

D-9 cats pulling sheep's-foot rollers, which compacted the dirt as it was being leveled and watered from two big 5,000-gallon Euclid water wagons.

It was a dangerous job working around all that equipment. I had to be on my toes to keep from being run over as I hurriedly walked up and down the fill, spotting, or signaling, the bottom dumps and big earth-movers, showing them where to unload according to the grade marks on the stakes on the shoulders of the road. I worked between the big rigs coming and going.

I rode to work with Carlos, another laborer from San Jacinto at first, but then drove the old Dodge when I took the job spotting for the bottom dumps and earthmovers. When I started my job as Dump Man, my hours were no longer the same and I put in a lot of overtime. My job changed from seven o'clock starting time to four in the morning and I worked until six or later in the evening.

I worked for Matich Construction from the early fall of 1959 until October, 1962. In the fall of '59, Jeanie and I took a bus to Santa Monica to shop for a car. I was told there were plenty of car lots there and I could buy a good used car that wouldn't cost me as much as at the lots in Riverside or San Bernardino. After looking in several lots I decided on a car we both liked. I spent most of the $1400 I had saved, and we drove home in a beautiful 1956 Cadillac Convertible. Jeanie loved the car. It was a real light Mountain Laurel pink, twin pipes with a black top and gold grill, It cost me $1300 out the door, I got a $500 discount because of a leaking transmission seal and the faded weathered top that needed to be replaced. We had to stop three times on the way home and put in transmission fluid. I didn't use the car for transportation to work.

Married in Tijuana

Jeanie and I were married in Tijuana Mexico in the evening of January 13, 1961. Tom and his girlfriend Madelyn went with us, and Tom was my best man. Jeanie and I were both worn out after we had our wedding party that night in a nightclub called the Trocadero. (What a coincidence. The bar had the same name as the place where I had danced with the little flamenco dancer on the Rock of Gibraltar. The little dancer was about the same size as my pretty wife.)

After we were married, I managed to get the oil leak fixed in our '56 Cadillac. Jeanie loved driving the big car; now it was hers. I started using her Ford as a work car after replacing the engine. The car ran good but overheated several times coming home from work and I intended to replace the thermostat over a weekend, and carried a five-gallon can of water in the trunk in the meantime, so I could refill the radiator when it boiled over.

One blistering afternoon, I had already stopped and put water in on the road coming from Redlands, but had to stop a second time on Highway 79. I spotted a turnout on the left side of the road near a small gravel pile and some bushes. I pulled the hood latch and sat there in the car waiting for the car to cool down enough to take off the radiator cap. After about ten minutes, I opened the door and stepped out and start-ed for the trunk to get the water can when I heard it–that rattler sound Dave and I heard when we lived near Double Adobe and had walked out through the mesquite to drive our milk cow home. I knew right away what that rattle was. That dry rattle, that warning. There was no mistakin' that sound. I stood real still trying to find out where that rattle was coming from. As I carefully looked at the big bush near the gravel pile, I saw it. The big diamondback came slithering toward the car.

I got a glimpse of it crawling under the car as I carefully went around the car to the passenger side, quickly opened the door and grabbed my gloves from the floorboard, a piece of re-bar and a long lathe with a piece of bright red ribbon tied to one end of it. They were my grade markers I used on the job. My plan was to kill it with the re-bar when it slithered out from under the car toward the pavement.

I stood with the re-bar poised and waited, waited for it crawl out from under the car. I kept waitin'. I knew it had gone under the car, but where did it go? I very carefully got back away from the car and looked under it.

No snake.

Wait a minute; did the thing somehow get up inside the car? But that can't be, the only place it could get up inside was the space where the brake and clutch pedals came up through the floorboard. I had circled the car and it was nowhere to be found. God only knows what the people in the passing cars thought that afternoon when they saw me out there wearing a hard hat and gloves walking around the car tapping the sides and fenders with the long lathe with the red ribbon, lathe in one hand and the re-bar in the other. I'm sure they thought I was a mental case, a real nut out in the middle of nowhere in the boiling heat, tap-ping on the sides of my car.

After pounding awhile, I was sure the snake was inside, maybe under the front seat or up around the radiator where it was nice and hot. Snakes did look for hot spots. After a half hour of probing and tapping,

I opened the trunk, took out the water can, set it in front of the car, and went to the driver side and eased the door open; still no snake in sight. I stuck the re-bar under the seat to try and rake the snake out in case it had gone under there. As I did, the re-bar touched an empty paper lunch bag, the rattle causing me to jump back and almost lose my balance.

Now that the inside was clear, I carefully unlatched the hood. After finding no snake, I unscrewed the radiator cap and filled the radiator, and just as I was screwing on the cap, I saw the snake cinching up around the left brake drum. I closed the hood and peeled out onto the highway. I could hear it thump, hitting the fender until it lost its grip. I pulled over on the shoulder, grabbed the re-bar and ran back and finished it off.

♪

Everything seemed to be working good for us until the late summer of 1960 when I had a near fatal accident coming home from work.

After finishing up the job spotting in Banning and Beaumont, I followed the job to San Bernardino to work on a section of Highway I-10 going from San Bernardino through Redlands where they were taking out big orange groves and replacing them with fill from the dry Santa Ana River to make the roadbed for the highway interchanges.

The job started at Waterman Avenue in San Bernardino and ended up outside Redlands. My work hours were the same as they had been, but I had to leave home earlier because of the driving distance. I had traded the old '41 Dodge for a '51 Dodge Wayfarer. The two-door sedan had decent upholstery, a strong engine, and the body was in good shape—no dents and good paint.

It was a Tuesday when I got off work at three o'clock and stopped off at a bar to relax and have a few cold beers while listening to some of the latest country music on the jukebox. I wasn't paying to attention to the time while sitting there having one cold beer after another. The sun was low in the sky when I left the bar. I knew I'd be late for my date with Jeanie if I didn't get on the road. The popular tune "Whispering Pines" was blarin' on the jukebox as I walked out the door. Jeanie got off work at five, and we had plans to go to Riverside for dinner and dancing at seven. It was after four when I left the bar to head home.

When I walked up to the car, I noticed the right front tire was flat. That was no problem. I had a good spare and the tools to change it in the trunk. I hadn't really checked out the spare when I bought the car, I had just popped the trunk and checked it out for tread, and I hadn't noticed that the tire was a bigger size than the others on the car. The rim was the same but the tire was larger. When I got the tire changed, I noticed the right front was just a little higher than the left, but the tire had plenty of air and good tread, so I wasn't worried about it. I put the jack and flat in the trunk and headed for home.

When I turned off on Highway 79 going toward Hemet, there was practically no traffic in either direc-tion, so I put the pedal to the floor, bringing the old car up to a speed far too fast for the narrow dipping road. I came down off a slight hill into a dip and when the car came out of the dip, both front wheels came off the ground, and when they landed hard on the pavement, the big tire on the right side hit the pavement first, causing it to fold underneath the car, making it impossible to steer, sending the car careening off the road. Hitting a bank on the shoulder of the road sent it end over end. The last thing I remember when I came down the hill was watching the speedometer bouncing between 80 and 90 and trying to hold the car on the road when the big tire on the right side hit the pavement. The highway patrol said the car went end over end three times before rolling six times on its side, coming to rest on its top twenty feet from a telephone pole where I lay beside the pole in agony with a shattered shoulder blade, four cracked ribs, and a fractured collar bone. My neck felt like it was broken.

To this day, looking back, things happened so fast I'm not sure if I was injured leaving the car and hitting the door as I was flung out or if I got my injuries from being thrown against the telephone poll. I'm pretty sure I got my shattered shoulder blade, messed up ribs and neck from being flung into the telephone pole. One thing for sure: I'm very lucky to be alive.

I was still lying beside the pole when the ambulance came. Jack, the owner of Jack and Flo's Bar, a bar at Lakeview where Tom, Jeanie, and I used to drink came along just after I rolled the car. He saw me lying there by the pole with the car lying on its top, wheels still spinning. He flagged down another car, telling them to send an ambulance while he waited there with me.

I was off work a month. No charges were filed against me for speeding. The wreck happened eighteen miles from Hemet. The day after the wreck, there was an article in the newspaper about it, and another one was published that same evening. Reports of the wreck were called in twenty minutes apart. The only mistake I found in the newspaper articles was the year of the Dodge I rolled. It was a 1951 Wayfarer, not a 1950.

Although I missed a month's work, I was able to go back to work at the same job. One Monday morning near the end of the job in San Bernardino, as I was waiting for the next bottom dump to spot, I overheard a conversation between two black laborers who were shoveling dirt around a culvert pipe. One was talking about his Saturday night at a bar in San Bernardino where a lot of black men went. The big heavy set one was telling his friend about the fight he had witnessed. He said two guys got into an argument and were walking toward the front door so they could finish their fight outside when the smaller man took out a razor and slashed the throat of the bigger man sending him staggering outside, spurting blood all over the floor and out onto the sidewalk.

He said "the jukebox was goin' full blast as the police was puttin' the handcuffs on the little man with the razor and loading the big man in the ambulance. The juke box was blarin' out the song "C. C. Rider, See what you have done." This tune was one of the latest blues tunes at the time.

1962

Before I was laid off in the fall of 1962, the foreman had told me that I was the first Spotter, the first Dump Man, who had ever worked for them that long without some kind of serious injury. He said that I wouldn't be out of work very long if I wanted to take a job working nights as a mechanic's helper. It would be a job servicing the equipment. I knew I would be making more than double the pay I was getting if I joined the Operating Engineers Union, and I would get the training needed to get an even higher paying job as a mechanic. I would start out working in a shop in Victorville, California.

The job was tempting, but I couldn't work nights that far away from Jeanie and our beautiful curly-haired, four-month-old boy with the big brown eyes, Jimmy.

Jeanie had saved a $1000 and wanted to move north to Oregon and try to find and buy an affordable home. She and I were both tired of living in the desert. She liked the Grants Pass area and had relatives living there, her cousin Glen, his wife Charlotte and their two children lived in the area, and her Uncle Lowell and his family lived north in Springfield.

Grants Pass reminded me of Williamsburg, Kentucky. It had two one-way bridges crossing the Rogue River, one entering town and the other leaving.

In late October, we rented a storage locker in Hemet and stored our furniture and things too big to go in the small trailer we rented to haul the rest of our things— clothes, bedding, and my musical instruments (my mandolin, five-stringed banjo, and a big, full-sized Epiphone bass fiddle. We loaded the small, covered rental trailer with enough groceries to last us several months, and loaded

Jeanie's big steamer trunk with all kinds of canned stuff, everything from canned cream and canned vegetables to cans of spam and potted meat, and we placed sacks of potatoes, beans and flour against the trunk to keep everything from sliding back against the tailgate.

I clamped on the rented hitch and hooked up the loaded trailer. We said our goodbyes to the family and left San Jacinto and headed north. Our first stop after crossing the border into Oregon was Medford off In-terstate 5 where we got gas. Our next stop was at Rogue River, a little town not far from Grants Pass, where we rented a cabin from Bill Gooley, a retired Navy man who was the owner. His office was in the main house, where he lived with his wife and teenaged son.

We didn't have much of a choice of cabins; they were all rented except for two that were close to the river. Jeanie didn't want to be that close to the river but Gooley knocked off five dollars on the rent and we stayed there for two weeks, and then used most of Jeanie's savings to buy a house off of Murphy Creek Road above the little town of Murphy. Murphy was near the Applegate River, seven miles out of Grants Pass.

We paid down on a two-bedroom house with two acres on Murphy Creek Road, about a mile from Mur-phy. The pie-shaped piece of property had been cleared of most of the timber, but our house and the house above it were backed by acres upon acres of timber that covered the mountains behind, extending into the Siskiyou Range, which stretches clear into Northern California, most of it government land.

The little house above us belonged to a nice old lady by the name of Minty Wilson. Minty, a relative of the man who sold us the house, was living with her daughter Jewel, but she was alone most of the day because Jewel worked in Grants Pass to support her mother, usually leaving for work early in the morning.

After being there a week, we found Minty to be a nice person but quite nosey. Jeanie had bought me three yellow sweatshirts that were on sale. Minty saw me working up at the barn on the hill one day and asked, "Is that the only sweatshirt you have?" And when she walked down the hill to get her mail, knowing Jeanie was in the house taking care of Jimmy, and that the neighbors, Mr. and Mrs. Reed, were working in the barn or feeding their cattle, she would kindly pick up our mail along with Mr. and Mrs. Reed's. She thought nothing of reading the postcards, or any mail that was already partially open that came from our families on her walk back up the hill. She would tell Jeanie and the Reeds: You got a postcard from so-and-so and they said they are all doing just fine.

One afternoon she asked me if I would wrap some outside water pipes for her to keep them from freezing. She said Jewel hadn't had time to insulate them. She said if I had the time, she also wanted me to get some stuff out of the attic for her.

After wrapping the pipes I got a ladder and unloaded the attic, setting it all on the ground in front of the house. As I was looking at the things before I carried them back inside, I noticed an old battery-powered Philco radio with hardly a scratch that was made in the 1930s. I told Minty I remembered us having a radio like that when I was a kid and I told her I'd like to have one like that myself. She said her youngest daughter had won the radio selling seeds before the war and died of an unexpected illness when she was just a teenager. I had just finished setting the things where she wanted them in different places in the little house and just as I turned to go, she said "Young man, before you go I want you to have this." As I turned, she handed me the old Philco.

We were buying the house and two acres for $5000. The house had belonged to a logger and it had no foundation except for some concrete blocks that were supporting some rotting timbers. The original house was a fifteen-by-twenty mill cabin that Minty's son had hauled from the mill yard in Murphy, and the add-on rooms and roof were a cobbled up mess. It had two rooms that were done in knotty pine, which Jeanie loved. The kitchen and the master bedroom both had knotty pine ceilings, walls and cabinets, and the cobbled up place had a fair-sized carport, a place to park the convertible out of the weather, and the property also had a small barn on the hill above the house.

We really loved it there, but the only jobs that were available in the area in and around Grants Pass didn't pay any more than my unemployment check. I would actually be making less money by the time I paid for my gas to and from work, so I did odd jobs working for the neighbors and started a charge account at the Murphy Creek Store, which I paid every month.

When it came to our Christmas Dinner, we certainly didn't have money for anything extra, anything other than the bare necessities. Our neighbors, Mr. and Mrs. Reed, who knew we were struggling to make ends meet, brought us delicious cream pies to have with our Christmas dinner. They were always willing to help us if we had trouble with the electrical system or the pump for our well.

One morning a couple of weeks after Christmas, I got up and was getting ready to go to Grants Pass to the unemployment office to check on work. I had been helping two nurses that were renting a house near us stack wood in the pouring rain over the weekend and was sick with bronchitis when I got up to get ready to go to town.

I went in the bathroom to take a hot shower and could get no water, hot or cold, so I got dressed and went to the little pump house down the hill to see why it wasn't working. When I opened the door, I could tell right away by the smell that the pump motor had burned out sometime during the night. The ground was frozen and I thought at first that the water in the pipes might be frozen and not letting the water through.

I walked back up to the house, grabbed the car keys, opened the trunk, got the necessary tools, removed the pump motor, put it in the trunk and went back in the house. Between coughs, I told her what had hap-pened to the pump and that I was taking it in to town to have it fixed. I then carried two five-gallon cans of water from the neighbor up the hill and got in the car to go to town. More good luck: the battery was dead. Mr. Reed had told me before that he had a battery charger if ever needed one. I loosened the battery clamps, and tapped on one that was a little tight. The whole side of the battery broke out spilling acid out under the car.

I took a bucket and carried water back from the neighbors' and washed off the mess. I brought a bucket and set it on the sink for Jeanie to have water while I went to town for a new battery. We were lucky to have such a good neighbor in those tough times. Mr. Reed took me to town, bought a new battery at the station he traded at, and took me to the shop to drop off the pump. I managed to pay him back by helping him fix his fence. Like the old sayin' goes, if I didn't have bad luck, I'd have no luck at all.

♪

We took trips to Springfield to visit Jeanie's Uncle Lowell, his wife Ann and their two sons, one of whom had caused a real tragedy in their family, a tragedy they were still mourning. He had accidently shot his older brother, the Lowells' oldest boy, when they were deer hunting two years earlier. He had bled to death before they could get him to the hospital.

On one of our visits, Jeanie's uncle took us on a tour of Salt Creek Falls. He was quite the fly-fisherman; he had fished in most of the streams in the area and he invited us to come up anytime and he would teach us how to fly-fish.

We enjoyed our drives, getting to see some of the beautiful country when I went searching for work. By the time April rolled around, Jeanie was pregnant with our second son. The baby wasn't due until December, but we were going to have to leave. I didn't have a steady, decent paying job, and no promise of getting one, and my unemployment checks would soon run out.

I had thought about putting in an application to go to work for the state police when I found out they were taking applications. After talking it over with Jeanie, we decided against it. She would have to spend too much time alone with little Jimmy while I attended the police academy.

I called home and told mother of our situation and Bill sent us $80. They were disappointed that we weren't coming back to Hemet when I told her that I planned to find work in northern California and later move back to Oregon.

Loss of Nuclear Submarine *Thresher* SSN 593

Before we left Oregon in 1963, the nuclear submarine *Thresher* SSN 593 sank off the New England coast on April tenth in 8,400 feet of water. It was being discussed on the news. The loss of that submarine caused me some concern because I was almost certain that I might have known some of the crewmembers. They were trying to figure out what may have caused the catastrophe. The boat was on a test run after leaving the Portsmouth Navy Yard. It made its fatal dive and imploded after plunging below its test depth.

It went down with all hands, two hundred and twenty miles east of Cape Cod, taking with it the lives 129 men, which included civilian yard workers along with the Navy crewmembers. The last message that came from the *Thresher* was heard aboard the U.S *Skylark* ASR-20, which was monitoring the test dive that fatal day. It was the garbled message, "Have reached test depth," meaning it had reached its maximum depth; then came the crunching sound of the bulkheads between the compartments collapsing as they imploded, sending the sub to the bottom.

In one of the newspaper articles I read it said that there was a possibility that the loss of control was due to a compensating water line that had split because of a faulty weld done when the sub was in the yard for repairs. They said the compensating water line might have split, spraying salt water on one of the main control panels causing the crew to lose control. That was the conclusion LCDR Dean L. Axene came up with after hearing the distorted messages recorded on the Skylark as the submarine made its final and fatal dive. LCDR Axene had been aboard the Thresher when it was first commissioned.

Axene was also my first skipper on the *Croaker*, a veteran of World War II, who had served on the USS *Parche* where he received the Bronze Star Medal for action against the enemy on *Parche*'s sixth war patrol in the mid-forties. He had also served as Executive Officer aboard our first nuclear submarine *Nautilus* SSN 571 from July 1954 until August 1955, and he also helped put the *Thresher* in commission.

By the end of April 1963, we had rented out the house in Oregon to the nurses living nearby so we would have enough to make the house payments, and we headed back to California. We didn't have the slightest idea of the problems we would have with the faraway rental a couple of years later.

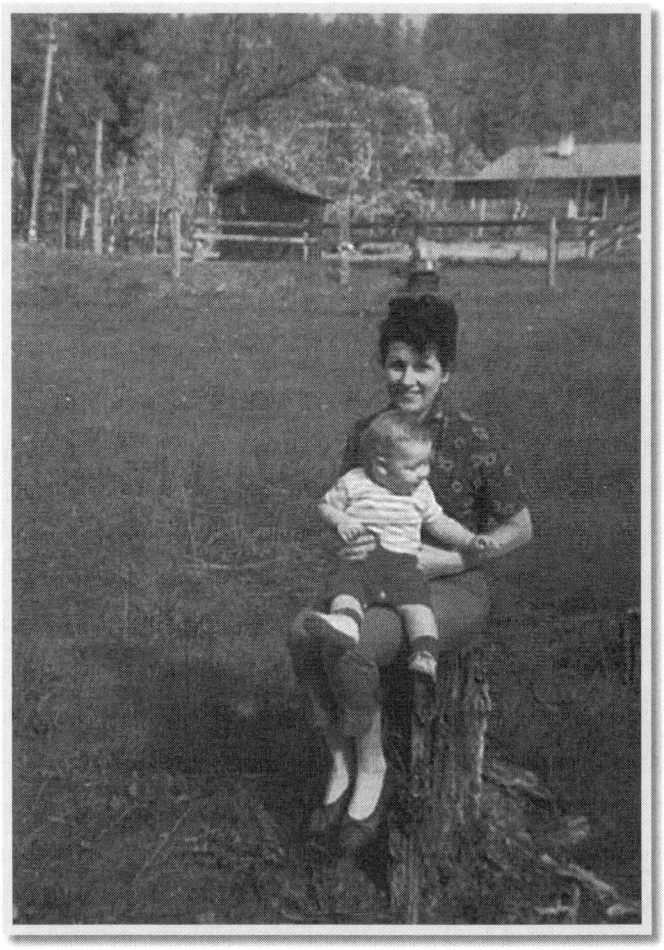

JAMES P. CREEKMORE

8 A GROWING FAMILY

Our first plan was to try to find a job in Crescent City just eighty-six miles away, or Eureka, farther down the coast over two hundred miles away, but Jeanie didn't like living that close to the ocean with all the wind and rain, so I called Weston Wright, a high school friend living in Santa Rosa, and he told me that there was all kinds of construction going on and I would have no trouble finding work. We went on south to Santa Rosa, which was 500 miles from Grants Pass. Our plan was to save enough money to replace the house on Murphy Creek Road and move back up there.

Our second little boy, Gary Lee Creekmore, was born December 30, 1963. Now we had two handsome little boys. I'll never ever forget those wonderful years. I can't say enough about those good years, those happy, fun years we spent together raising our sons, and later enjoying them, their wives and our three beautiful grandchildren.

I often think how fortunate I was to have such a precious and caring wife and family. We were just happy being together and were grateful for what little we had. Jeanie wanted to go back to work. She had put in an application to go to work at a grocery store, Safeway, and with her experience working for AJ Bayless in Arizona, she was sure to get a job as a cashier or meat wrapper, but after a sleepless night I talked her out of it. I told her when we married that she would never have to work again as long as I was able to make a way for the family.

Jeanie managed our money very well. Keeping up with or get-ting ahead of the Joneses just wasn't our thing. I often think back of the chances I took in my younger years while growing up, and some of the things I did both in civilian life and during the time spent in the Navy, and realize how lucky I am to have made it this long. Going through submarine training, through the various stages to qualify, could have caused me to have serious health problems down the road; things I did then, I never gave a second thought. I know most of us look back and wonder if we would have done better had we chosen a different path.

Santa Rosa

When we first came to Santa Rosa in April 1963, I checked in with the Labor Union list and was working three days later. I went to work as a carpenters' helper, pouring foundations and carrying lumber and stacking bundles of cedar and asphalt shingles on rooftops of the tract houses being built in Bennett Valley, a section in east Santa Rosa.

We had moved out of the little cabin on Redwood Highway South and were living back on Mendocino

Avenue where we had stayed with the Wrights. I was at that time also working on the Cadillac at night. I had pulled the engine and put it in the shop and had it rebuilt, and ended up pulling the engine a second time because the man who did the work failed to re-place one of the plugs at the end of the camshaft causing a total loss of oil pressure.

I was upset, but I didn't call him or do anything to add to his misery when I found out that he was under a lot of stress and had been going back and forth to see his wife who was in the hospital in San Francisco. The poor man's wife was dying of terminal cancer.

As I was getting the engine back together, I lost a fitting for the new oil filter I installed before having to take the engine out the second time. The filter, called a Franz filter, was a new design that hadn't been put on the market. The big, stainless Franz filter was still in the experimental stage and the regular parts houses had no inventory of these parts. There were few people in our area that sold the Franz filters and parts for them. I was given the address of one such person who might have what I needed. I followed the directions I was given and found the big house on Walker Avenue where the guy who sold the filters was supposed to live.

I'll never forget that mistake, and I'll certainly never forget my first and lasting impression when I met Owen Duane Nunnemaker, or I should say "Slim," as he was known, the man who had just recently moved into the big house on Walker Avenue in the southwestern outskirts of Santa Rosa.

I pulled up in the driveway, stepped up on the porch, knocked on the door and stood there waiting. A few minutes later a voice rang out "I'll be right there."

I waited patiently as a tall, skinny, hunched-over man came to the door in a trail of smoke with a freshly lit cigarette between his long, tobacco-stained fingers. He was wearing black slacks; a once white, but now yellowed, tobacco-stained shirt; a string tie; and long, narrow half-shined black shoes.

What caught my attention right off were the long fingernails on his skinny right hand as he took long drags off the cigarette clamped between his fingers. The nails were curved and looked more like talons than fingernails. In all my travels, I had never seen a man with fingernails that long; women yes, but a man, no way. His facial features and the way he was dressed reminded me of a character from a horror movie. He looked like the undertaker in Count Dracula or Burnt Offer-ings. He could easily have played the part.

After we introduced ourselves, we stood on the porch and talked. He said he had just moved in a week ago and had no idea where the man who lived in the house before him had moved, the man who sold parts for Franz oil filters. I told him I was looking for a Franz filter part and mentioned that I sometimes worked on cars after work and on weekends. He said he had been a carpenter and was drawing disability and on pain pills and was suffering from a bad back injury after falling from a ladder on a construction job some months be-fore.

He said that so far, with his disability checks and the money he was making in the stock market, he was doing okay, but like the most of us, he could always use a little extra cash.

He was quite a talker, quite a storyteller. As we talked I found out that his wife Dee did upholstery work for extra money. He said maybe we could get together and trade some mechanic work for upholstery work like new seat covers for my pickup or car. He said he and his wife would soon be moving to a big house in Sebastopol where they would be running a home for emotionally troubled children, but he would keep in touch. He smiled, showing his long, yellow, piano-key teeth as we shook hands and I stepped off the porch. I had no idea what an evil person the man that I had just met really was.

In the fall of '63, I worked on some of Nunnemaker's cars in exchange for cash and some very professional upholstery work done by his wife Dee. Dee upholstered my pickup seat with a nice, heavy, black and white, Naugahyde cover.

Nunnemaker had family in Ukiah, a small town about seventy-five miles north of Santa Rosa. The first job I did for him was to tow his big Chrysler 300 with the big bombsight taillights from the family pear orchard in Ukiah. I repaired the engine in our backyard on Mendocino Avenue and he started using it for his main transportation. After getting the big four-door Chrysler back on the road, I rebuilt the engine in his 1957 Plymouth station wagon. I was also doing small welding jobs for him and others to help out with the family income.

One Saturday afternoon he and his wife stopped by the house to pick up an antique rocking chair, one that I had repaired by welding a broken spring. They had with them one of the young, emotionally disturbed children they were taking care of.

The little eight-year-old boy couldn't stand still and right away Jeannie and I could see that he was getting on Slim's nerves. Slim scowled at him and yelled. "Settle down, Spook." Jeannie glanced over at me and rolled her eyes as Slim yelled over and over again for "Spook" to settle down. Spook–of all things to call a mentally retarded, emotionally upset little boy.

Of all people to be having the responsibility of caring for emotionally disturbed children, the state couldn't have chosen a more warped and cruel person for the job.

Years later, in 1974, Slim had just come back from the islands in the Caribbean and he called me one evening to say he wanted to share with me some pictures he had taken while on his trip. We had stepson Bill watch our kids Jimmy and Gary who were teenagers at the time for a few hours while we visited the Nunnemakers in Sebastopol.

Slim had a nice German Luger that he had shown me before and we had arranged to trade that for a German-made revolv-er that I had, a .22 caliber with a spare cylinder that held .22 magnums. The magnum's shells were longer and held a much bigger powder charge than the regular .22. I brought it with me that night, but he backed out of the trade.

Slim had the screen all set up when we got there. He was anxious to show us pictures of the forty-foot boat he traveled on between the islands and pictures of him standing under the big palm trees with his arms around the pretty women he had been with at the different ports he visited. His wife brought us cold drinks as we sat down to watch the show. We noticed right away that Dee wasn't pleased with his show of pictures. She wasn't happy at all.

As he was getting set to show us more pictures, I got up and went to the restroom that was down the hall in the big house. I walked about halfway down the hall toward the restroom when I saw the little boy, the one who had walked in front of the projector. He stood leaning against the wall with his legs spread, standing on the tips of his toes, his arms also spread with just the tips of his fingers keeping him from smashing his head into the wall. He was spread-eagled like some criminal waiting to be searched.

When I got back to the living room where Slim was showing his pictures, I asked him what was going on with the little boy in the hall. He said he was doing "fingers and toes," a punishment that the children received if they were rude or unruly.

He finished showing the pictures and we sat and talked about the islands. Slim talked about the money that could be made if he owned a boat to haul food supplies and machinery to the different ports. He knew I'd been in the Navy and had some experience with small boats. He said he would find a nice place for us to live on one of the islands and I could come down and run his boat.

Before we left, I had to make a quick trip to the restroom to relieve myself of some of the beer I'd been drinking. I couldn't believe it. There in the hall stood the little boy doing his best trying to keep his spread-eagle position.

When I came back from the restroom, I told Slim he should let the boy get some rest. He didn't say anything; he just looked away as Dee walked us to the door. On our way home, Jeannie and I talked about how he had treated that little boy and wondered if the state was aware of how those kids were being treated. We decided to have as little to do with him as possible from then on.

He called me one afternoon in the middle of May that year and asked me if he could store some stuff in one of my stor-age sheds. He said Dee had kicked him out and he needed a place to store some of his furniture. I told him that we didn't want to get involved in any of his family squabbles.

Near the end of that month, he stopped at a little bar on the corner of Gravenstein Highway and Bloomfield Road. After a few drinks, as he got up to leave, he showed the bartender his Luger, the one he was supposed to trade for mine, and said he was going to put a couple notches in the handle. He walked out and drove up to the big house off of Bloomfield Road, rang the doorbell, and shot his wife through the heart. He stepped over the body and grabbed the phone, tearing it from the wall as the kids and house assistants ran for cover. He then ran up behind the house and was hiding in the orchard when the police arrested him.

Now we knew exactly what a cruel and evil man he was: Owen Duane Nunnemaker was sent to prison for the murder of his wife Dee (her real name was Alice) on May 31, 1974.

♪

Our first house, or I should say cabin, was off Old Redwood Highway south of Santa Rosa. We lived there for a couple of months while I worked as a laborer on a housing project in Bennett Valley.

The little moldy cabin, one of the three cabins in the small court owned and managed by an old Scotsman was where we lived until my friend Weston Wright went back to Oregon to take a teaching job in Harper. Weston had spoken to his land-lord before he moved, and we were able to move into the house after he left. While we were still living in the cabin on Old Redwood Highway, I started having engine trouble with our '56 Cadillac. After checking it out, I found it had a burnt valve and a water leak around one of the head bolts. I re-moved the cylinder head but had no one to give me a ride to take it Joyce machine shop, which was two miles away.

There was no one at home in either one of the other cabins, and the old Scotchman running the place was no help, so I wrapped the heavy cylinder head in newspaper, put it in a gunny sack, threw it over my shoulder and carried it to the machine shop, having to stop and sit down on the side of the road several times to rest.

The machine shop put in a new valve and seat and bored out one of the head bolt holes for an oversize bolt to fix the water leak. I got home, started to put it back on and discovered they had drilled out the wrong bolt hole, so I carried it back and had them do it over again. I was dog-tired that evening when I finally got it all back together. That evening I had Jeanie come out and help me get it started. At the time she was about seven months pregnant with our second little boy Gary. I was bent over the fender getting ready to adjust the carburetor after she got it started.

I didn't know it, but the line coming from the fuel pump wasn't tight enough and it sprayed gas all over the engine causing it to catch fire. I ran over by the house, grabbed a pan and tried to get water from the faucet between the two cabins, but the old Scotchman had it turned down to a trickle. We were very lucky. The man in the next cabin saw the blaze and came running out with an old blanket that we threw over the engine, which smothered out the fire.

We were glad to get moved into the little house off Mendo-cino Avenue. It worked out good for us. It wasn't far from my job as a carpenter's helper when I started working on the new jail and courthouse. We were living there when our second little boy Gary was born in Santa Rosa Memorial Hospital. Gary, another beautiful little boy was born on December 30, 1963. He was another cute little boy, a little brother for Jimmy to grow up with.

I was still looking for year around work and I needed steady job instead of the uncertain construction jobs to support my family. While working the construction jobs I took after coming to Santa Rosa, we had finally started to have a little spending money. We bought a television and washing machine. We didn't keep the old washing machine long before I upgraded and bought a new one to replace the old, or I should say "antique," Maytag with wringers. The old washer was like the one our folks used in the thirties and forties. The original small gas motor was replaced by an electric motor. It did the job, but I traded it in on a new one because I was always afraid Jeanie might get her hand caught between the old hard rubber wringers, which could have caused serious injury.

The spring months of 1964 were busy but worrisome months for us. My real worry was Jeanie's health. She had been going back and forth to the doctor since Gary's birth in December.

The doctor had put her on birth control pills, pills that almost took her life. She ended up back in the hospital with a blood clot in her right leg.

Late one afternoon when I came home from work, she was sitting with an ice pack on her right leg. When I saw how swollen it was I loaded her and the kids in the car and rushed her to the hospital. I called mother in San Jacinto as soon as I got Jeanie settled in the hospital. Mother came up from San Jacinto the next day to help take care of the boys.

Jeanie was critical for several days as the doctors were trying to dissolve the big blood clot before it strung out into her lung or worse yet, into the blood vessels in her brain. To say the least, I spent some sleepless nights. I could never have handled it if Mother hadn't been there taking care of the kids and giving me moral support.

I wasn't only lying awake nights worrying about her, but worrying too about my next steady job. I have to say that the years 1963 and '64 were tough years for us. It seemed there was always something to worry about. If it wasn't a problem collecting rent money for the old house we had rented out in Oregon, or the car breaking down, it was something else happening to cause real concern, especially for a worrywart like me.

Moving to Moorland Avenue

In 1964 we moved out of the little, cramped house on Men-docino Avenue in north Santa Rosa to a much nicer and much bigger house south of town. The place on Moorland Avenue was owned by one of the carpenters I worked with while working for Guy F. Atkinson. He had bought two houses on an acre and a half. He was remodeling and living in the larger of the two, making it into a two-story. We rented the second house and really liked the place. What a big difference from the little cabin we had been living in! A house with a big living room, three bedrooms, two bathrooms and a big yard was a nice place in the country to raise our little boys. It was the nicest place either one of us had ever lived in.

After I finished my labor job with Atkinson, I worked one more construction job for Dinnwidde Construction, working on the big Emporium building going up in the new shopping mall in Santa Rosa. Before I was laid off, Atkinson offered me a job as Labor Foreman in Oakland working on the coli-seum. I turned it down because it was too far to drive, and I wasn't about to move my family to a city of that size with its high crime rate and all its killings going on.

I had another application in with Schoonmaker, a company in Sausalito about forty miles away. It was a big shop that re-paired and rebuilt large diesel engines for ships and power plants. Some of the plants were shipped to the islands in the Caribbean and other parts of the world.

I had experience working on the big General Motors V-16 278-A and Fairbanks Morris opposed 10-cylinder diesels with the double crankshafts, those used in World War II in diesel electric submarines.

The shop manager who did the hiring was going to the Me-morial Day races and told me to call him when he came back from Indianapolis; he would let me know when to come to work.

By then I had managed to buy a half-ton 1961 G.M.C. V-6 pickup, with an eight-foot bed and a Gem Top canopy. We had made a trip to San Jacinto and hauled back to Santa Rosa what little furniture we had in storage there.

On the night of the twenty-fifth, I hardly slept; I had two job offers. I tossed and turned all night trying to make up my mind about which job to take. The following day was the day I was to call the foreman at Schoonmaker if I wanted the job.

It was also the day I was to start work as a Package Car Driver for United Parcel Service. When I left the house that morning, I had made up my mind. I was going to take the job at Schoonmaker in Sausalito. The job doing mechanic work was the job for me, instead of jumping in and out of a package car for UPS.

The week before I had passed the UPS driving test and had gone through three days of training to learn the proper meth-ods expected of me as a delivery driver. I was being trained to take the route that started in Healdsburg, a small town north of Santa Rosa, and the route included Cloverdale, farther north.

The first day out started pretty good. I was shown how to properly sheet the packages, copying the numbers off the packages from different companies, some had to have signa-tures and some had release numbers; those with release num-bers had to be placed where the person receiving them could find them, but carefully hidden from those walking down the street in front of the house.

When I stopped to make a delivery, I pulled the key, unlocked the bulkhead door between the cab and the packages, pulled the package from the shelf, shut and locked the bulkhead door and stepped out of the delivery car. I hurried up to the door, writing down the numbers off the package as I walked. If no one answered the door, I was to go to the neighbor for a signature.

I had no trouble with the first few stops. The route was actu-ally a good place to train, a more or less rural route with no heavy traffic. About the fifth stop my trainer and I got a real surprise. I had just come back from ringing the doorbell with no response. I walked around the front of truck where the guy training me stood, and he started to walk across the street to get a signature from a neighbor working in his front yard. All of a sudden, there came an ear-splitting scream and a loud clap of hands. My clipboard and package went straight up in the air and came crashing down on the pavement in front of me. A tall teenager jumped out from behind the truck, jumping up and down and screaming and clapping his hands as a big German Shepherd came running toward him from behind the house.

The man who had signed for the package said the teenager was retarded and told us he did that every time he called his dog.

The rest of the route went pretty smooth, but when I came home that night after finishing the training, I wasn't sure I really wanted the job.

I left early the next morning for the UPS Center. Jeanie had told me it was my choice and she wanted me make the deci-sion. By the time I walked into the building, I had made up my mind. The job in Sausalito would be the one for me.

When I arrived at UPS, I was plenty early and most of the drivers hadn't clocked in yet. The manager, O.T. Randolph, was just entering the office. I went into the office and told him that I had changed my mind and wasn't going to take the job. He was surprised and felt let down. I apologized and we shook hands and I hurried out to the pickup and headed south toward Sausalito.

The manager at Schoonmaker had told me to be sure and give him a call when he got back, so I pulled off the highway to the nearest phone booth and made the call.

After telling me about the good time he had at the races, he told me to load in my tools and come on down. I stood there silent for a minute before I answered. I had very few tools and nothing big enough to work on those big engines, and I didn't have money to buy them, pay rent and make payments on my pickup, too. I told him I would have to buy a few more tools and I would call him in couple of weeks when I got the tools.

My head was spinning. Now I had no job and no promise of a job so I drove back to the UPS Center just as the trucks were pulling out the door and hurried over to the office to talk to the manager. I asked him if he had called the city yet to tell them I had turned down the job.

He said he hadn't. I told him I had thought it over and would take the job. He told me the truck was loaded and waiting, handed me a clipboard and the necessary paperwork that went with the route, shook my hand and said to give him a call if I had any problems.

My job with UPS started May 26, 1966. Randolph was one of the best managers. He was liked and respected by all the drivers. The job not only paid well, but had extra benefits as well. It was a good job for a family man.

Before that, I had even tried to go to work at San Quentin at one point, but l failed the test. After hearing from Mike Foley, a friend that quit UPS and went to work at San Quentin, I was thankful that I hadn't been hired. Mike had been working a part-time job on weekends for the Sonoma County Sherriff's Department before he went to work at San Quentin. He was doing well on his job until he was attacked by an inmate and beaten down with a lead pipe, which left him with serious injuries that would impact the rest of his life.

Financial Relief

Jeanie and I were now living better and having more than either one of us had ever dreamed of. No more layoffs because of bad weather or holidays. I actually made more money over the holidays, especially before Christmas, than any other time of the year. I worked all the overtime hours I could get. We no longer were living from paycheck to paycheck, and were able to enroll both the boys in the Santa Rosa Christian School, while my stepson Bill, a teenager, attended public school.

We were able to give them many of the things we had never enjoyed while growing up. Jimmy and Gary shared the second biggest bedroom, and Bill had the other bedroom in the front of the house. We even had enough money to buy things for the kids and gas money to go places on weekends.

When I bought a carpet and some nice new furniture for the big living room, the boys watched in awe as men placed a nice big couch and reclining chair on the bright new carpet. As the men were getting back in the truck to leave, Jimmy looked up at me so seriously and asked in low voice, almost a whisper, "Dad, are we rich?"

We bought two young billy goats for Jimmy and Gary. I got a good buy on them from a UPS customer, Mrs. Carstensen, who had a goat ranch on Rainsville Road near Petaluma.

The boys named the two billy goats Abendigo, Jimmy's goat, and Shadrack, Gary's. The goats hadn't been weaned from the nanny and had to be bottle fed. They really had fun feeding them and playing with them; but as they begin to grow, they became harder and harder to keep the pen. The boys were saddened when we sold them back to Mrs. Carstensen, but the sadness disappeared when I took the money and bought them each a Red Ryder BB Gun.

I had weekends off and we would go to the coast so the boys could fly their kites, fish, and catch crabs. On most of those trips, when the boys were first learning to cast, I spent the biggest part of my time untangling fishing lines and changing reels. On one of those trips, one of the dairymen I knew from my route showed us a little pond up behind his dairy. He said the pond had never been fished and probably held some pretty big catfish.

The only fish caught out of that pond was the one big catfish Jimmy caught. It was a boneless that he caught on a five-pound line that I helped him carefully drag from the muddy water. We threw that one out because it smelled like the stag-nant water from the pond.

Sometimes we would go to Guerneville and go hiking in the redwoods or up above Cazadero in the hills where Jimmy and Gary could shoot their BB guns and check out the pools in Cazadero Creek. I took them to the places I knew they would enjoy, many places that were on my delivery route. I took them to dairy farms that had fish ponds where there were no game wardens to worry about.

Once we had been fishing at Bodega Bay when a game war-den walked up on us and wrote tickets for Jeanie and the boys—fishing without a license. We were lucky there were no fines; they just had to buy a fishing license. After that, Jeanie always had a handy whistle in her tackle box for warning the boys if the game warden was near if they were fishing in a place they shouldn't be or catching tiny crabs.

Some of our weekends we took trips to San Francisco to Fisherman's Wharf and visited places like The Wax Museum or spent a Saturday afternoon at Golden Gate Park. All of this was within driving distance from home.

One Sunday we packed our ice chest and food box in the pickup and headed for Annapolis, a little wide spot in the road northeast of Sea Ranch off Coast Highway 1. Jeanie and the boys had never been there and knowing how Jeanie loved the mountains, I knew she and the boys would enjoy the trip.

Everything was going as planned. There wasn't much to see in Annapolis, just a store and post office. The road going in from Sea Ranch wasn't nearly as steep and winding as the one going south coming out at Skaggs Springs Road. While driving back, we decided to stop and stretch our legs and let the boys play in the shallow creek underneath the bridge where the road comes out on Skaggs Springs Road, a steep, narrow, winding road that runs between Healdsburg and Highway 1. The road winds around through the mountains and comes out at the Stewart's Point Store. overlooking the ocean on the coast.

After letting the boys play in the creek to work off some of their restless energy, we enjoyed a few snacks and sodas, put the ice chest up against the cab in the back, and got back in the pickup. The four of us sharing the front seat, we pulled out onto the narrow, winding road and headed for Stewart's Point.

As we came down the steep hill into the sharp curves, I would press on the brake, trying to keep as far to the shoulder as possible to keep from getting hit by someone coming up the hill on the narrow, almost one-way road. We were doing fine, taking our time coming off the hill, when I started to brake before going into a tight curve and the pedal went all the way to the floor. The '61 GMCs hadn't come out with a separate braking system for front and rear, so if a brake line broke or leaked, it would bleed off fluid from both front and back and you would have no brakes left except the emergency brake, or trying to gear down to come to a stop.

I pulled the emergency brake on and off to slow our speed and I managed to grind the transmission into low gear as we smoked and ground our way across the highway in front of the Stewart's Point Store. We breathed a sigh of relief when we finally came to a stop.

If I hadn't replaced the original three-speed with a bigger transmission some months before, coming off that hill could have been a real disaster. I was surprised I didn't wipe out the transmission when I ground the gears and jammed it into first. Jeanie went in the store and called my stepson Bill and told him about the problem while I crawled under the pickup and found the problem. The flexible line going to the left front brake cylinder had split, spraying the brake fluid that was dripping from the floorboard underneath. The master cylinder was bone dry. The sun was getting low when stepson Bill and his friend brought the brake line to the stranded truck, and the new line he had brought didn't fit. I managed to plug off the brass fitting and stop the leak. Then I had Jeanie and the boys go back home with Bill while I drove the pickup home with three wheel brakes and a growling, grinding transmission.

I wasn't about to let them ride back with me. Not with only three wheel brakes down that winding coast highway. The next week I replaced the brake hose and the broken cluster gear in the transmission.

♪

The population of Santa Rosa and the outlying areas had really begun to grow by the mid-seventies. New businesses and tract homes were going up everywhere, and it seemed like there was a bank or fast food place on every corner. We didn't have to leave town for a special piece of furniture or appliance. We had it all right here in downtown Santa Rosa or in one of the stores in the malls.

To some it was great, but for us, we didn't like it. There were too many buildings going up too fast and too much traffic. We didn't want to live in a city. I had put in for a transfer to Oregon, but found out that the only one available was a part-time job starting in November in Eugene as a car washer and part-time mechanic. I turned it down.

We couldn't afford to move without the promise of a steady job. With the boys in private school and very little money in our savings, we couldn't take a chance on moving. We didn't want lose what little we had worked so hard to gain. Besides, when we moved we wanted to have a piece of land as big as the one in Santa Rosa, at least a half acre.

Our decision to stay worked out for the best. Our sons were able to finish high school in Santa Rosa and Jeanie didn't have to go to work as she probably would have had we made the move.

Santa Rosa had become home to us. We didn't actually hate living here. Jeanie and the boys liked our little farm, the little half-acre plus. The boys liked to go fishing out at the coast and they liked fishing in the ponds at some of ranches where I delivered packages. They also enjoyed the goldfish we had in an aquarium for them in the house.

One Saturday, we all got together and made a little fish pond outside to put some of their goldfish in. Both of the boys had helped me pour the sidewalk around the side of the house and the patio slab, where it still stands today with both their initials drawn in it. With their help, I cut and placed the reinforcing screen in the bottom and on the sides of the pond. I shoveled in the cement while Jeanie and the "little guys" used their bare hands and shaped a nice little goldfish pond near the house.

No one can deny who the builders were because the pond has their fingerprints all around the rim at the top. A couple of years later, we decided to move the pond. I dug it out, pried it up, scooted it onto several pipes and rolled it to the back. It now sits near the fence behind the shop under the shade of a pussy willow tree Jeanie had planted some years before. This part of Moorland Avenue then was a good place to live.

We liked the convenience of being able to take one-day trips to go on hikes in Cazadero, go to the coast on fishing trips, or check out Armstrong Grove, which has the big redwoods, in Guerneville. Our place was also less than a two-hour drive from San Francisco, if we wanted to go sightseeing.

The climate too is hard to beat. We do get frost and heavy rains in the winter months. Back in the 60s and 70s, the heavy rains were our biggest problem. When we had a frost, we had no trouble saving the outside plants, we always covered them; but the heavy rains were a different matter. Because of the small culvert pipes and lack of storm drains, the street would flood and the water that backed up from those storm drains flooded both our yards, front and back.

Both yards were lower than the ground on which the house sits. Our first winters were rough when the heavy rains came. The boys had to wade through a foot of water to the street to catch the school bus. The chicken and goose pens sat on high ground; but, to get to the pens we put on rubber boots. Win-ter storms with record rainfall can be pretty rough on those living in this part of California. The town of Guerneville, which sits on the banks of the Russian River, is one of the worst places to be during the flood season. When the heavy rains come, roads and bridges wash out and mudslides cover many homes, and many of the small houses have been carried away and end up in the sand at the mouth of the river in Jenner.

Not only those living in low-lying areas have problems. Some of the ranches in the hills that are near rivers and streams pre-sent a real challenge to get to during those floods. I got some real experience driving the flooded back roads for UPS, and some stops were impossible. I put on a lot of miles just back-ing out of the flooded narrow, roads in daylight, and it was really rough backing up in the dark. The UPS truck had no back-up lights and the only way to light up the roadway was to use the four-way flashers. Many times during the heavy floods, the customer's package or packages would have to be left with a neighbor or nearest grocery store.

One such flood ended up in tragedy at Frank Ambrosini's ranch. One of my best remembered customers was Frank Ambrosini. Frank lived not far from Tomales on his family's old homestead place, on the land his parents had bought in the 1920s. He liked getting his special tractor parts and cata-logue orders shipped UPS; that way, he didn't have to waste time making trips to town. UPS brought them right to his door.

Several times when I made deliveries, he would give me a big paper bag filled with potatoes, more than enough to last us a month.

One day while making a delivery, I followed him inside a barn close to the house. The walls inside the weathered barn were covered with all kinds of posters from the late 20s and early 30s. The colorful metal posters used to advertise different produce and products that were sold during the Great De-pression were all pretty much in mint condition. None had been exposed to the weather outside the old barn.

Frank's farm was really hit hard by this particular storm. He had just come in from checking out his flooded potato fields when he noticed that the long bridge that crossed the Estero had washed away and the pens holding his geese had blown over. The geese had flown away looking for shelter from the storm.

He was lucky the original pair he had started with were still in their pen near the house, but he was overwhelmed at the sight of all his losses, all he had worked so hard for all those years, all now going down the Estero.

In his depressed state, he shot and killed his faithful dog and tried to take his own life by shooting himself in the head. The damage to his property and the loss of the large flock of Canadian geese he had raised was apparently more than he could handle. He was taken to the hospital but only lived a couple of weeks. What a tragedy for him and his cousins whom I knew from my Petaluma delivery route.

The fertile eggs of the original pair that started his flock, the ones I had delivered in the second-day air package to him about three years before. were shipped UPS.

When some of the first had hatched, I had told Jeanie of the flock he was starting and took her and Mother out to see them. Frank was Mother's age and they had a lot in common to talk about since Mother too had been raised on the family homestead on Becks Creek.

Before we left that day, I tried with no luck to buy a pair of geese for our little farm. He said it was against the law to sell them. He thanked us for stopping by, wished us well, and gave us a nice bag of potatoes.

Two years after the big storm, I was making a delivery to Diekmann's Store in Tomales; I had a package for Bill, the owner. His sister said he was in the back and I was directed to the yard behind the store where Bill was working on his car.

There in the yard near the back of the store sat a pen that held Frank's original pair of geese, plus two more. I asked if the geese in the pen were some of Frank Ambrosini's flock. He said, "They sure are. They aren't for sale, but I'd like to get rid of the two younger ones. If you happen to know someone who would like to have them, let me know." I couldn't believe I was so lucky. I came out that Saturday and picked them up. They weren't a pair, but Jeanie had her Canadian geese. She finally had the special birds she had wanted for so long.

We made a small pond in the pen for the Canadian geese and a larger one for Jeanie's other geese and mallard ducks. One day, when Jeanie was going out to feed, she heard the geese all sounding off and noticed that several honkers from Frank's flock had landed in the pen as they flew over. She had seen them fly over before, but had no idea any of them would land. They were squawking and trying to get in the small pen.

Some evenings as the sun is going down, they still sound off as they fly over the house after leaving the pond north of Rohnert Park on their way toward the coast or one of the ponds between here and the coast. From what I've read, they mate for life and some live to be twenty years old.

9 MY TALENTED WIFE JEANIE

Jeanie was so talented; she wasn't only a talented dancer, but was also was good at ice-skating and roller-skating. When we settled in Santa Rosa, she started ice-skating again at the new ice arena built by Charles Shultz, the famous cartoonist who had lived in the area much of his life. She had just started to get used to taking the boys skating with her when one afternoon, having just put on her skates and stepped out on the ice, she was struck from behind by a kid learning to skate. They both ended up on the ice. I was in tears that night when I came home and saw her broken wrist. In her younger years, she had always dreamed of being a figure skater. She was a natural on skates. The young skater wasn't hurt, but Jeannie ended up with a cracked bone in her wrist. She never skated again after that; the boys and I sure hated to see her give it up.

She was a natural farm girl, a natural at gardening and flower growing. She was like Mother; as the old saying goes, she had that "green thumb." Every time we went on trips, she would take a cutting from a plant or tree and keep wrapped in a damp cloth or can of water until we got home, and then plant it in a pot, and nine times out of ten it would grow.

She also raised different breeds of chickens and some of the finest geese, both domestic and wild. She had her own recipes for smoking fish and turkeys, and making jerky. The jerky she made was the real stuff, nothing like the sugar sweet stuff you buy at the store that is so loaded in brown sugar, you can hardly taste the flavor of the beef. Her smoked salmon and albacore were the best.

After being smoked, the albacore was put in half-pint jars, put in the oven, heated and then sealed. She was very professional when she canned or cooked. We never got tired of her delicious fish dinners. She had so many recipes for frying and baking fish. When I went fishing on Jeanie Two, and brought home sacks of fish we would be up half of the night getting it ready for the smoker and freezer. Our little family enjoyed the wonderful home canned food. When I had a good catch when I took the boat out on weekends, I took some of her deliciously smoked fish to work and gave to the guys I worked with, and sold dozens of eggs for a dollar a dozen to Doug Runcimun, one of the guys in that worked in management, from Terre Haute, Indiana. He was a farm boy that was raised on fresh farm eggs.

Jeanie took the boys to the Sonoma County Fair those years that she raised such beautiful gardens. I went several times on weekends but didn't go nearly as much as they did.

As I look back on those wonderful years, I know that she would have won first prize with some of the plants and flowers she raised and especially the beautiful jars of canned fish and vegetables and fruits. The jars of pickled peppers, cherry peppers, and green beans were perfectly placed in pint and quart jars.

She had every kind of recipe for cooking meals served in all different parts of the world and dozens and dozens of her own recipes that she kept secret.

A caring wife and dedicated mother to our two sons, she was more than I'd ever dreamed of. She had no problem when I worked late, which was most of the time (even more late hours during the holiday season). She never thought twice about me sending Mother much-needed money to help my brother Bill pay their bills and keep food on the table, or to make improvements on her house, an old house she had bought to replace the small trailer they had been living in.

Along with the money orders, she usually sent mother a box of cherry chocolates and a pair of shorts or slacks. As I've mentioned many times before, to my family and me, Jeanie was special.

At night when I came home from work she would have my supper waiting in the oven, and even though she was tired and had spent the biggest part of her day working in the garden, in between taking the boys to have swimming lessons or to some school function, she always looked her best. She was so pretty. When she served me my late suppers, although many times she would be "bone tired," she never had that haggard look. Sick or well, she looked as though she had just come from the beauty salon.

She always kept my clothes nicely ironed and hung in the closet with any torn or frayed part neatly sewn and any missing buttons had been replaced. My underwear, shorts and t-shirts were always carefully folded and put in my part of the dresser. And I would find the things she would buy me in a neat stack on the bed. It was like she read my mind, when I would think of something I needed, I would find it spread out on the bed that evening when I came home from work. She was equally as thoughtful while we were raising our precious sons. Their clothes were always clean and neatly pressed when they went to school. She always confronted the teachers and the principal if she thought they were being mistreated in any way.

I'm living alone now and I think of those words Mother used to say to me, "You could have looked the world over and never found a wife as precious and thoughtful and caring as Jeanie." Believe me, I couldn't agree more.

Like all married couples we had our little spats. The few ar-guments we had weren't shouting matches using filthy lan-guage like the foul-mouthed society of today. The arguments we had were behind closed doors, not to be heard by the chil-dren. One thing that did really make her mad was the Friday nights I stopped to have "a couple of beers" after cashing my check. I would call her after work on Friday nights to tell her I would be home in a little bit, after a couple beers with the guys. Usually, instead of being home before nine, I would get home at ten thirty or later after spending twenty dollars of the family check, often the twenty dollars we needed to pay our bills. It didn't take her long to put a stop to my Friday night "couple of beers":

One Friday night I came home about eleven thirty. When I pulled in the driveway I noticed the car was gone. As soon as I stepped up on the porch, I noticed the door was ajar, I opened it and walked into the living room, and I panicked. The furniture was in disarray, the coffee table lay on its side against the couch, and one of the table lamps on one of the end tables nearest the door lay on its side, hanging like there had been some kind of fight or struggle.

I called out to Jeanie and the boys and no one answered. My heart sank. I ran into the back yard calling them. I was just getting ready to call the police when Jeannie and the boys came in the back door. She told me right then and there that she and the boys would be gone for sure if I kept up what I was doing. From then on, when I got my check on Friday morning, I stopped by the house and dropped off the check or she met me at the UPS center or on route to a pick-up.

She really taught me a lesson. Not only did I come home earlier, but I also began saving the family money. After that, I always had a few dollars to buy beer after work if I really wanted to. After the scare that Friday night, I never came home late.

Many Friday nights after that, she let me invite some of the guys I worked with to come to the house for snacks and beer or dinner. Everyone who had dinner at our house really en-joyed her delicious food. Her fish and chips were made from the fish I caught while fishing aboard the Jeanie II. She had her secret beer batter recipes for frying fish. The beer batter recipe was a family favorite. The big platters of fish were served with delicious salads made from vegetables and tomatoes fresh from her garden. Anyone who sat down to her delicious snacks or suppers couldn't believe what a fantastic cook she was.

The Story of *Jeanie* II

In the summer of 1970, I was making a delivery to a sail-maker on Magnolia Street in Petaluma when I saw a boat. George Goodell, the sail-maker who owned it worked in Sausalito and had brought the twenty-three-foot Cruisalong Cabin Cruiser (made by Kris Kraft) home and had it up on blocks in his front yard, He had planned on doing some minor repairs on the engine and repainting it before taking it back to the boat yard in Sausalito with the intention of selling it.

I had made several deliveries to him before and had briefly talked to him about his job as a sail maker. Two weeks after first seeing the boat, I had another package for him. I asked him how he was doing with the boat. He said that he was too busy to work on it and asked me if I knew of anyone that might be interested in buying it, just as it was. He said he wanted $1000 for it "as is." I told him I would love to buy it myself, but couldn't come up with that much cash. He said if I really wanted it, we could work something out, and I said I would talk to my wife about it, and let him know the next time I saw him.

Jeanie and I talked it over and we decided I could start paying for it with the extra money I was making doing welding jobs and the money I was making putting a roof on a big, two-story Victorian house on Old Lakeville highway, about five miles from Petaluma.

None of my regular paycheck would be used to buy the boat.

The owners of the big Victorian I was working on were Mr. and Mrs. Goldie. They were both quite wealthy. Mr. Goldie was an auto broker in San Francisco and

Mrs. Goldie was related to the Rainiers who owned the Rainier Brewing Company in Washington. It was almost like she had money to burn.

Not only did I roof the Victorian, but I also put new shingles on a forty-foot pool house, a converted chicken house that covered their big swimming pool. She didn't like the color of the new green shingles already covering the roof, so she had hired me to cover them over with shingles to match the ones I had put on the Victorian.

One Saturday after finishing the job on the pool house roof, I was in the living room of the big Victorian discussing with Mrs. Goldie what she had planned for me to work on when I came out the following weekend.

Before I left, I asked her if she and Mr. Goldie knew much about Mare Island. I knew that they had lived in the Bay area for quite some time. I ask her if they had been aboard some of the Submarines there. She said it had been some time since they had been there and didn't know the names of the subma-rines that would be there now. She said if I liked to read about the submarines of World War II she had just the book for me.

She walked over to the big bookcase and pulled out a big thick book and hand it to me. The book titled Submarine Operations of World War II Written by Theodore Roscoe. I thumbed through the book and was amazed at all the authentic pictures and logs written by the Skippers aboard the different submarines while on patrol in the late 30s from the beginning to the end of World War II.

I had been aboard several of the Submarines mentioned in the book in New London. The diesel powered ones, some that had survived the war and some that were named after some of those lost in the War and got to go aboard the first nuclear boat then.

I've never been aboard any of the big nuclear missile subma-rines, those bigger than the Nautilus, but I'm sure the crews are more comfortable in their confined space than the crews on the Nautilus and the diesel-powered boats.

The Silent Service as they were called didn't get nearly the news coverage as the Surface Vessels because of the secrecy of their operations.

After Jeanie saw how much I enjoyed the book, she surprised me with a copy, what she said was my early Christmas present.

It wasn't long after I bought the boat for $100 down and a $100 a month. I rented a boat trailer in Sausalito, brought the boat home and put it up on blocks in the front yard and worked on it at nights and weekends, caulking it, and painting it. I also built a trailer so I wouldn't have to pay rent on slip to birth it in Bodega Bay.

The original name I discovered when I peeled the paint off the transom just above the brass boarding steps in the center of the stern. I uncovered the faded name Suva, named after one of the Fiji Islands. The original owner may have been a Navy Officer stationed there during the Korean War.

I renamed the boat *Jeanie II* and had planned on having a bright red rose painted under her name, like the big bright rose on the City of Portland, the train I had ridden coming west from Chicago; but like a lot of things I wanted, I never had that extra money.

The boat turned out to be a lot of fun but like all wooden boats it took a lot of work keeping it up. The wooden boat should have been left in a birth, in the water, not on a trailer.

♪

Jeanie's brother Bud, his wife Shirley, and their family lived in South San Francisco and often came up for weekend visits; his family enjoyed having plenty of space to run and play. The only bad part of their visiting was the fact that they were both heavy smokers, smoking one cigarette after another. And they smoked in the house, leaving us both with splitting headaches and a house full of second-hand smoke. Back then it seemed like ninety-nine percent of people were smokers and thought nothing of smoking in your home; they would have been insulted or mad if you suggested they smoke outside.

We took trips to their place on some of the holidays or other special occasions. One of those special occasions was a birth-day party given by one of Shirley's uncles for one of her un-cles and her grandmother, "Momma Martinelli".. Shirley wanted us to meet her family, so the party was part birthday party and part family reunion.

The party took place, not in the dining room of the small house; but in a large, dimly lit room in the basement . It took a few minutes to get adjusted to the semi-dark surroundings. The room got its light from big candles spaced out on two, long, folding tables. The party started off with big spaghetti feed by candlelight with plenty of red wine and sourdough garlic bread.

In all my years to this day (and that's a lot of years), I can't remember a time when or where I'd made a bigger ass of myself or been more embarrassed as I was at that party.

It took awhile for the spaghetti to arrive, so in the meantime, we enjoyed our wine and garlic bread and got acquainted while we waited. I hadn't eaten anything since breakfast, before we had even left home, and after about three big glasses of wine, I was, shall I say, a little tipsy?

The big, smoke-filled basement had the clamor of a bar with the crowed all chattering, laughing and puffing on cigarettes. I was seated next to Momma Martinelli, with Jeanie and our two little boys to the right of me. We filled our plates and began to wind the spaghetti round our forks. I had just finished a bite of garlic bread and washed it down with a big gulp of wine when I turned to tell Mama Martinelli what a delicious birthday dinner we were having.

One of the ladies had just placed a nice plate of spaghetti in front of her. She turned her head and I started to speak. I noticed a hair on her upper lip. curving into her mouth. My first thought was that nice old lady has a hair in her food; I can't sit here and let her put a hair in her mouth and gag and maybe throw up her the first bite of her birthday dinner. No way was I going to sit by and watch this happen to this nice old lady. I thought she was about to take it down with a big fork of spaghetti. Just as she started to take her first big mouthful, I cleared my throat and said, "Momma Martinelli, just a minute," and reached over to pluck the hair before it got mixed in with her food. She pulled her fork away as I moved my hand toward her mouth, getting a firm grip on the hair between my thumb and forefinger. As I gently attempted to lift the hair to keep it from going into her mouth, Momma Martinelli's upper lip pulled up with it.

It was attached, growing out on her upper lip. Oh My God, I thought, what a stupid ass I'd made of myself.

I cleared my throat and turned back around, as Jeanie was pressing hard on my leg and kicking my foot, letting me know what she had just seen.

Mister Nice Guy had screwed up big time. I was lucky every-body was busy talking, sipping wine, and stuffing their faces with garlic bread and spaghetti and most didn't see it. Thank God for the dim candlelight. Looking straight ahead, I took a gulp of wine from my glass and started stuffing down my food as if nothing had happened. I was too embarrassed to say another word to Mama Martinelli. What could I say?

On our way home that night, we talked about how funny it really was for me playing the role of Mr. Nice Guy, and having something as ridiculous as that happen.

♪

As much as I would have liked, I never did try to buy another boat. The only mementos I have left of the *Jeanie II* other than pictures are the original compass, the brass parts of the fold-down helmsman seat, and the little round wooden seat, which I made into a tall stool for my shop.

It seemed like every time we got caught up on our bills and started having some fun, something would come up to take away from that fun. One of those "distractions" putting it mildly, was a feud with a neighbor over a fence line at our place on Murphy Creek Road. The renters were late with the rent money most of the time, but we managed to keep the old place rented.

The trouble started in the summer of '75. I had a three-week vacation and planned on stopping and camping out along the Rogue River, while the boys and I replaced the old broken down fence on the hill. The fence separated our two-and-a-half acres from the Kavedre's property, the last house on a dead-end road that went on up into the woods to a 160-acre parcel purchased by a Mr. Mackatee six months earlier. At that time, the neighbors held meetings that I was unable to attend, protesting the sale of the 160 wooded acres. They were concerned about the acreage being logged and divided up into housing tracts.

Jeanie had paid down on a nice piece of property in the Idaho pan handle a year before, and our plans were to camp out and spend a week putting up a fence on the Murphy property and then go north and enjoy the last two weeks of my vacation camping out on the 14 acres we were buying in Good Grief, Idaho.

The 160 acres of timber on the mountain off Murphy Creek Road had sold and the access road to our property was the road that ran between our property and our neighbor's, the Kavedre's. Jeanie and I wanted to get a fence up to keep the owners of the 160 acres from cutting through our property below the fence line which we thought might cause washouts when the heavy rains set in.

As we started putting railroad ties in as anchor posts every twenty four feet to support the 317 feet of horse wire and metal posts we were putting in, Mrs. Kavedre, the neighbor's wife, started giving me a bad time over the property line as we were putting up the fence. She said that we were fencing off her property. In other words, I was cheating, giving myself more land than I had a legal right to. She could plainly see that I was following the old fence line that was actually (we discovered after having it surveyed) giving her a foot of our property. I even showed her the metal survey stake I was going by where the road splitting the two properties started from the main road that came up the hill. What a lash-up. We ignored her and put up the fence, dividing the two properties. When we drove down the hill toward our campsite that last evening, we were very proud of our accomplishment. The nice, tight fence was in place. No more problems with boundary lines. We had no idea of the nightmare that was yet to come concerning that fence.

We arrived at the Good Grief campground late in the evening, our first time ever going that far north in the state of Idaho on Highway 95.

We had only seen pictures of the 14 acres Jeanie had bar-gained for through a real estate offer she had found in the Bonners Ferry Newspaper. The paper was sent to me from one of my UPS customers who had moved to Bonners Ferry the previous year. We had the opportunity to buy 30 acres, but we couldn't afford the joining 16-acre parcel so we settled for the 14 acres. The property was a real buy. We had no idea it was so heavily wooded. We were quite pleased with the property and its location. It had 500-foot frontage on Highway 95 and was four miles from Eastport, the immigration station on the Canadian border.

We had all our camping gear, including a nice summer tent that was plenty big for the four of us. We had thought about camping out on our own property, but after checking it out, we changed our minds.

The next morning after breakfast we parked our pickup near the highway and climbed the hill going up into heavily wood-ed property. We had just come to the remains of what had once been a small log cabin when I looked down at a boulder that had been overturned.

The big rock had large claw marks on it made by a good-sized bear that had overturned it to feed on the grub worms beneath. I had a loaded .9-mm Luger in a holster on my belt; but shooting a big bear with a handgun with a bore that small would only make it mad. It would be like hitting it with a rock.

We cautiously walked back to the pickup. It was just too dangerous at that time to walk 14 heavily wooded acres with a large bear, and god-only-knows what else that might be in there, looking for food. I wasn't about to take that chance. We returned to the campsite, spent the night in our tent and decided to go into Canada the next morning. I left the Luger with the camp manager, and crossed the border at Eastport. We went as far north as Fort Steele.

We really enjoyed our trip home. We went east from Sandpoint to Thompson Falls, and then north again to see historical Flat Head Lake, and to fish the Flat Head River. Jimmy and Gary and I were using spinners and lures with no luck, without a single bite. Jeanie really showed us up when she caught twenty-three small cutthroat trout, baiting her hook with whole kernel yellow corn that she got from our camping supplies. Nearly every time she cast out she pulled in a fish, and caught even caught more fish when we stopped to take some pictures of an old sawmill sawdust burner not far from Missoula.

The burner, sitting out in the dried up flat, had been made to look like a Dutch mill. The dome on top that had once been a spark screen had been covered over, still keeping the dome shape. About five feet down from the top the windmill, a shaft came out in the center of four wide blades. There were two levels of rooms in the building. The windows were framed with a short canopy at the top, and the lower level had a narrow porch with a handrail that went all the way around. There was a circular pond nearby and Jeanie and the boys caught a few more trout. We went from there to Missoula and headed home. It was good to get home and sleep in a nice comfortable bed and not in a cot in the tent or the bed of the pickup.

I don't think a summer passed that we didn't go on camping trips, on long holidays or spend our vacations away from San-ta Rosa, especially after I gained seniority and could choose the times I wanted off. Like I mentioned earlier, working as a package car driver for U.P.S was never a boring job; every day was a new experience. My workdays went by fast, and it was hard to keep from running late if I took time to explain anything concerning our service. There definitely was no spare time to visit with the customers. We were trained to ring the doorbell, get that signature, hurry back to the delivery car, and rush on to the next stop.

Several weeks after we came back from vacation, I received a phone call from our renter on Murphy Creek Road. She said the nice fence we had put in was now a tangled up mess of wire, metal posts and railroad ties. The renter had the sheriff come out, and she said she was sure the Kavedres were the guilty party, but without a witness, the sheriff couldn't do anything. Jeanie and I were furious. We couldn't imagine anyone knocking down the whole 317 feet of fence, Maybe the railroad tie on the corner, but the whole fence? No.

That night I didn't sleep a wink, thinking of what I'd have to do to fix the problem. I just had to go back up there and deal with it. I knew Milo Kavedre was the guilty party, but proving he did it would be another matter.

I had recently bought a '76 Ford F-250 pickup and taking a trip to Oregon. traveling the winding highways that threaded their way up and down the mountains. would be a good jour-ney to check the truck out. When I explained the fight over the fence on my Oregon property to Randolph at U.P.S., he let me have the time off, using a couple sick days to go up and take care of the matter, so, that Friday night I headed out.

I arrived at Grants Pass Saturday afternoon, rented a motel room and drove on out to Murphy. I stopped first at the house and talked to the renters and looked over the messed up fence. It must have taken some real work to bring it down. The wire laid all tangled up on my side of the fence line facing down the hill, the metal posts and railroad ties had all been pulled and thrown down on top of the tangled wire. There were chain marks on the railroad ties a foot up from where they went in the ground where a choker chain had been cinched for pulling them out. The more I looked, the more enraged I became.

It was just getting dark when the Kavedre's came home. As I drove up the hill, the sound of Jeanie's voice kept going over and over in my head.

"Now don't go up there and get in a fight with those people. You lose your temper you might end up getting' killed or killin' somebody and end up in jail."

I parked my pickup in the renter's driveway and walked up the hill and pounded on the door. Milo Kavedre's wife came to the door and asked me what I wanted. I told her I wanted to speak to Milo about the fence. She said he wasn't feeling good and was lying down and told me to come back the next day and we could talk. I was tremblin' mad as I drove down the hill and headed for Grants Pass. I kept thinking over and over how I would get even with the S.O B.

I had just started over the one-way bridge going back to the motel, when a car darted in front of me where the left lane merged into the right as the road narrowed down crossing the bridge. As the car swerved to get in front of me, it struck my front fender and bumper, bending the bumper and left fender, and breaking the headlight. The guy driving the car was given a reckless driving ticket and his insurance paid for the damage when I got back to Santa Rosa and had it fixed.

I called Jeanie as soon as I got back to my motel room to let her know about the fence and the wreck. She told me again not to lose my temper and fight with Kavedre, I could easily end up in jail.

The next day when I went back up the hill to Kavedre's, and I pulled in beside a new four-wheel drive Chevy pickup. I could hear grumbling inside as I pounded on the door. His wife came out and Milo stood just inside wearing glasses. When I had met him just after he bought the house from the former owners, the Olsons, he wasn't wearing glasses. I guess he thought that the glasses might somehow keep him out of a fight.

I asked him why in the hell he tore out my fence, and he said he had backed into it and damaged the back bumper on his pickup. He said he wasn't driving a pile of junk like I was. I stepped off the porch and screamed, "Come down here in the yard, you son of a bitch, and I'll see just what you're made of." As I again started for the door, he said in a nervous voice," We can settle this without a fight. I'll have it surveyed, and I'll put it back nice and tight just like it was." It was all I could do to keep from going in the house after him.

The next day I went to an attorney for some legal advice on the problem and was told to put posts back in to mark the boundary and post no trespassing signs on them, and then have the renter get in touch with me if anything happened before the property was surveyed.

I went back out, put in the posts with no trespassing signs, called Jeanie, and headed home. I was home about three weeks when my renter called again with more bad news about the fence. She said the posts had been pulled out again. I phoned the attorney to tell him what had happened. He said he had no idea Milo Kavedre was the person who was taking out the fence. He said there was a "conflict of interest" and he could no longer handle my case. In other words, he had also been advising Kavedre to keep the fence down to gain the property through "adverse possession," a law that grants the property to the person or persons continually using it. He was telling him to take down the fence while at the same time telling me to put it back up.

I finally got it settled by getting a lawyer in Santa Rosa and turning Mr. Byrd, the Grants Pass lawyer in to the Bar Association. After paying for a survey shortly after that, Gary and I went back up and put the anchor posts back in once again.

We kept the place rented until we sold it in 1992. We had all kinds of renters, including Jeanie's dad, who I helped move from the Oregon coast in 1967. He was only there a year before his death in the winter of '68. Upon his death, Jeanie contacted his brother Lowell and his nephew Glen and told them about it. She asked them to notify us about when and where they were going to have the funeral.

We received no immediate answer. None of the family seemed too interested. We could find no kind of insurance, other than her dad's railroad pension, and we had little time or money to deal with the situation. We had our two little boys with us, and I was taking time off from work to deal with the problem.

We decided not to pay the funeral home to hold the body. We took it on our own and had the body cremated, which some of the family didn't approve of. I stayed with the boys in the waiting room of the funeral home while they held the funeral. We never did find out what happened to the old man's money. Jeanie said that he had no faith in banks. We had searched the house for any kind of papers on a savings account or any cash he may have had before the funeral. We thought he surely would have saved some money, a retired railroad man who lived alone all those years.

The only thing we found in the way of money was a few coins in change in an old worn-out suitcase in the bedroom closet. He had either buried what savings he had somewhere on the property, or the "good neighbor," Mr. Reed had "handled it" for him. To this day, we have no idea what happened to his money, if there was any.

The Tragic Years

It was like the old expression Mother used to say about bad things happening from day to day, "When it rains it pours." Once these bad things started to happen, it seemed they would never end. It's hard to put in words how these how these tragedies affected the family. The years from 1977 to 1980 were some of the roughest of my life, for me and for our extended families. Those were sad years. I know a lot of people out there have gone through the same things my family did, if not worse.

Death and dying is a natural thing, and death is expected for all of us. We wonder why good moral people and innocent little children suffer and pass away in pain and misery while murderers, child molesters, and hardened criminals live long lives and die of old age. We don't usually find answers.

The rough years for us started just before Christmas in 1977 when Jimmy's dog Red, a fine purebred Irish Setter, died. We had taken him hiking up near Warm Springs Dam and then to the coast for a walk along the beach where he became so excited seeing all the different kinds of birds going in all different directions, not knowing which ones to follow. When we got home that night, we couldn't get him to eat. It was really unusual for him not to touch his food. We took him to the vet the next day, and a substitute for our vet suggested that we leave him there for observation; the regular vet would find out what his problem was when he came back after Christmas.

Poor Red didn't live but one night after we took him in. When the autopsy was done, the vet found that Red's intestines had knotted up from all the excitement of chasing the birds at the coast. The loss of Red was really rough on Jimmy and Gary. Jeanie and I were crying right along with them after we got word that he had died.

After Red's death, Tom called us and told us he was on his way to Portland to bring Lila back home to Arizona. Her se-cond husband Johnny Johnson, a big rig driver, had shot him-self to death after an argument with his brother and some other relatives over his dad's will.

Johnny had come home the night of the argument, gone in the bedroom and killed himself as Lila was fixing their supper.

Not too much before that, they had made a brief stop at our place on their way north to Portland. They had a truckload of produce and couldn't stay very long. We didn't know that it would be the last time we would see Johnny and Lila together.

A short time later we had another death in the family. This time it was Dave's oldest daughter Theresa. She died in a snow-covered field about sixty miles from Burns. The car she was riding in with friends after a drinking party slid off a country road into a deep ditch filled with freezing water and slush. She and her friends, all soaking wet and freezing, were trying to make it to a ranch house about a half mile away when Theresa stumbled and fell face down in a puddle of slush, sucking up slush and mud from the pasture. When the rest of the party reached the ranch house, they began calling relatives and friends in Burns to send an ambulance or some-one out to bring them home. When they reached the ranch house, they were all so loaded–trying to make phone calls, trying to get dry and warm–that they didn't notice that Theresa wasn't among them An hour later, her sister Leah and a state policeman arrived. Leah had found Theresa lying in the snow and slush and noticed she was having trouble breathing. She tried to revive her but it was too late; she died before they could get her to the hospital.

We were all devastated by her tragic death, especially Leah who had tried so hard to save her. Theresa was so young, so talented. It was a real shock to all of us, but more was yet to come.

♪

The next year, summer of 1978, brought another heartbreaking death in the family; another one of my nieces, Sandra, the youngest of Bill's twin daughters, drowned. She was with her husband, her brother-in-law and her mother-in-law when the accident happened.

Sandy had never learned to swim and was afraid to be anywhere near deep water. When the family went on outings, the kids always played in the shallow water in a stream or at the edge of a lake.

The day before Sandra went to visit her in-laws, she had a quarrel with Pamela, her twin sister, Sandra decided she needed to get away for the weekend. She and her husband decided to go fishing with her brother-in-law, Pam's husband and her mother-in-law. The twins, having married brothers, shared the same mother-in-law, of course.

They had started home from the fishing trip just before dark and the pickup they were traveling in got stuck in the mud at the end of a bridge they were crossing. No matter how hard they tried, they couldn't get the pickup out of the deep mud hole. While the brothers went to locate someone to tow them out, Sandy fell in the stream and drowned.

Not knowing how to swim, Sandy slipped and fell into the rushing muddy water stream when she was trying to help push the pickup. Her body wasn't recovered until the next day. Why they didn't search for her the whole night through, we'll never know. After they got the pickup out, the three went home.

The mother-in-law said the last time she saw her, she was sitting on the small bridge waiting for her husband and brother-in-law to return with a tow truck.

No one notified Bill and the family until the next day. Poor Pam, she had no idea when Sandy left and went to the in-laws after they had their argument that she would never see her identical twin, whom she loved so much, alive again.

The accident had been reported to the local sheriff who was a relative of the in-laws; but, it was never really checked out. My twin sisters, Laura and Leta went to see Sandy in the mortuary and wondered what she had hit when she fell in to make a big bruise and puncture mark in the top of her head.

I was told that Sandy had had heated arguments with her mother-in-law before. Maybe the two had gotten into a fight. I don't think it's really clear of what happened that fateful night. As much as I wanted to, I wasn't able to make it to her funeral.

Two years later, her dad, my brother Bill, passed away. He had been having heart problems before Sandy died, and was so grief-stricken after her death, it made it impossible for him to get any rest. Her death really took its toll on all of us, but especially her dad. It kept him awake night after night. It was just too much for his already weakened heart. He had just finished grocery shopping when he had a massive heart attack beside his car in the grocery store parking lot and died on the way to the hospital.

I took the earliest flight I could out of San Francisco and flew back to Tallahassee where the family picked me up. We drove north to Havana where the funeral was to be held on the twelfth of May. Mother and the family felt bad for me when they realized that my brother was being buried on my birthday.

The night before the funeral, Pamela and I decided on the closing song at the gravesite. The name of the song we decid-ed on was "If." The song was beautifully sung by one of Pam-ela's friends, a young man in his late twenties. Bill's death brought real sadness to us all. He was only 56, and I had just turned 46. And we were all worried that Mother, then in her late seventies, would be the next one to go. The recent family tragedies were always on her mind, and Laura was really concerned about her; we all worried about her. But, with Laura's loving, caring help, she fooled us all and lived to be over ninety.

♪

When my brother Bill had started having heart trouble in the late '70s, I took two weeks of my UPS vacations and flew back to Florida to help Mother with the things that needed to be done on the upkeep of the house while Jeanie and the boys stayed home to take care of the birds and animals.

We had talked about having a burglar alarm but never got around to it. Jeanie told me not to worry about anyone trying to break in at night while they were asleep. If one of the dogs didn't wake her, she readied the poor people's burglar alarm by hanging pots and pans on the doorknobs and piling them in front of the doors. Jeanie, being raised in poverty like I was, had never lived in a house with a burglar alarm. While I was away she always slept in the room with the boys and always slept within reach of a knife or loaded pistol, which I had taught her to use if someone were to break in.

She was little, but didn't tolerate any kind of mistreatment of our boys. If she was told by one of the boys that he was being treated unfairly in school she took him to school the next day and confronted the teacher who had given him a bad time and also had a talk with the principal letting them know she wouldn't tolerate such behavior.

She certainly wasn't afraid to speak her piece, whether it was to a teacher or car dealer.

When we traded for our second, used Cadillac before Christmas in 1967, I found out that they had sold us the car knowing it had a broken piston ring. I had her take it back to them because I couldn't take time off from work. The next Saturday when I had a talk with the mechanic who repaired the car, he said that she had to be held back at the office door to keep her from punching the dealer in the nose when she found out the dealer had already sold her nice '56 convertible.

She wasn't a troublemaker, but wasn't afraid to fight. I found that out early in our marriage when she hit one of the guys I was fighting with in the head with a beer mug. She was little, but you better watch out if you made her mad because she would clobber you with anything she could get her hands on.

I sold the Jeanie II and her trailer in 1981 for $5000 and sent the money to Mother in Florida to help her pay for a house, which she moved onto the property to replace the mobile home she had been living in before the death of my oldest brother.

I had taken a week off to fly down to Bill's funeral. He was buried on my forty-seventh birthday on May 12, 1980. I well remember the funeral and how broken-hearted the family was. Bill had been through so much heartache and stress over the loss of his daughter Sandy. Mother had me keep his pistol, a Ruger .44 magnum, to be given to a family member or grandchild sometime in the future.

9 ON THE ROAD WITH JEANIE

In the summer of 1988, Jeanie and I decided to drive to the East Coast. I wanted to show her as many places as I could, some places I'd been and some places I hadn't. We planned to go to New London where I would take her on a tour aboard the *Croaker*. Two years earlier, much to our surprise, I had found an article and picture of the After Torpedo Room of the Croaker while thumbing through one of the many Readers Digest books we received. The book was titled "America from the Road."

Jeanie had belonged to the Readers Digest Book Club for several years and we had quite a selection of books, everything from "Discovering America's Past" to "The World's Last Mysteries." We had just about anything you would want to read about this country's history. In the book "America from the Road," the one and only picture of a submarine in the article was, of all things, a picture of the After Torpedo Room, my sleeping quarters while I served aboard the *Croaker*.

This was quite a surprise to me as I thought the old boat had been scrapped. I had mentioned it to Bob Vanderlen, one of the guys I worked with, and he said he was going back that way on his vacation and he would stop in New London and check it out. He would bring me back some pictures. When he arrived in New London and went to the pier where it had been tied up alongside the Nautilus, the Croaker was no longer there. He was told that it had been moved to a "Mothball Fleet" in Portsmouth, Virginia.

When vacation time rolled around, I decided not to split up my vacation, spending half of it Florida working on mother's house and the other half with Jeanie and the boys. I had four week's vacation plus five days of unused sick days. That would give us plenty of time to drive to Portsmouth and back down the coast, and stop and spend time with Mother on the way back.

We talked it over and decided to take Jeanie's Bronco and not the pickup. It got better mileage than the pickup and as small as she was, we could both easily sleep in the car when the back seat was folded down. We could stop at campground along the way and save money that would otherwise be spent staying in motels.

Driving back east would give us a chance to see some of the historic places we both had always wanted to see. I had passed through some of these places before but never had time to really see them.

I got my vacation pay and we paid my stepson Bill and his two children to feed and water Jeanie's birds and water her plants. We loaded the Bronco with our sleeping bags, flashlights, a Coleman lantern, and an ice chest, and headed out.

Our first stop of interest was The Great Salt Lake and the tabernacle where we took a few pictures. After leaving there, we took Highway 80 into Wyoming and stopped at Fort Bridger. The fort was named after the famous Jim Bridger, and they had quite a display of firearms, some used by the famous hunter, fur trader and scout, including the famous Gatling gun. Jim Bridger was famous for helping the early pioneers going west over the Oregon Trail in the late1840s and '60s and also for scouting for the military in the late 1860s.

After leaving Fort Bridger, we traveled east as far as Cheyenne and headed south to Denver. We planned to go from there to Cripple Creek where I would get to show Jeanie where I had lived in 1937 when I was just a little over four years old. But we had second thoughts when we realized our California Ford Bron-co with an engine smothered in smog equipment might give us problems in the high altitudes of Pikes Peak and Cripple Creek. We couldn't afford to take the chance, so we drove on to Aurora.

After spending the night east of Aurora, we headed for Kansas City, about six hundred miles east on Highway 70. We stopped at several little towns on the way. One of the towns I wanted Jeanie to see was WaKeeney, Kansas, a small out-of-the-way place about the size of Burns, Oregon.

I had stopped there on my way to visit Mother in the late 70s when I rode a Greyhound bus from Santa Rosa, California to Havana, Florida, a most memorable bus ride across the States. It was the first time I was ever on a trip where the vehicle developed engine trouble and broke down in a remote spot, forcing us to leave the hot bus and wait for two hours in the sweltering heat. We left the bus, walked across the highway and bunched up in the shade of a scrawny tree off the side of the road. We waited there for an-other bus to come out from Omaha to pick us up.

Before the breakdown, I had enjoyed a beer break while we had a thirty-minute layover in WaKeeney. Me and a couple other passengers went to a bar close by and had a couple of cold beers with pretzels and cheese and string cheese, the first I had ever eaten it.

I got acquainted with several interesting travelers, one was a man from the Soviet Union who boarded the bus in Salt Lake City. He spoke good English and the miles clicked away as we talked. The conversations I had with him were very enlightening. He was very straightforward in answering the many questions I asked. He filled me in on some of the latest things happening over there and told me about the strict security of the Russian borders.

He said all cars and trucks were thoroughly searched with dogs and armed police at the borders, and there were soldiers with guard dogs on both sides of the tracks. All the trains and trucks that crossed the border going either way were made to slowly pass over huge, lighted mirrors to spot anyone clinging to the underside of the railroad cars. This made it impossible for anyone trying to cross the border without showing legal papers. He told me of thousands of people living in poverty and said that in the cities, the hotels, eating places and other places that had flush toilets, you weren't allowed to flush the toilet paper. The used toilet paper was saved. The restrooms had recycling baskets for all the paper goods.

After seeing enough of WaKeeney, Jeanie and I headed for Kansas City, our next stop. Our plans were to stop in Leavenworth and see the federal prison north of Kansas City.

Our brief stop at Leavenworth was a shaky experience to say the least. As we were driving by the prison, we wanted to get as close as we could so I could take close-ups of the guard towers. I pulled off the main road that went by the prison onto a dead-end street that went down a hill behind the prison, a perfect place to get some good pictures. We pulled down the hill and parked. This was great, there were no people and the small parking lot had only two cars in it with nothing to block my view.

We were so interested, looking at the prison from the main road, we had no idea that the road we took going down the hill was off limits to tourists, and we were parked in a spot that was off limits to the public.

I had just snapped a picture and was getting ready to take another when two armed guards in uniform stepped out of a side door of the prison just below the tower and started walking toward me with their hands on their sidearm.

Here I was, standing there with a camera in my hand, dressed in dirty, faded jeans and a work shirt, looking pretty grubby with two days growth of beard, and Jeanie sitting in the car, a car with out-of-state plates and a loaded .357 magnum under the seat. We both thought we were in deep trouble. But, that was our lucky day. We both breathed easy when they simply told us that no cameras, visitors, or tourists were allowed in that area. They said I could take all the pictures I wanted to from the main road. I got back in the car and we left. We were lucky they didn't search the car. If they had, god-only-knows what kind trouble I would have gotten us into. We were lucky that we both didn't end up behind bars.

From Kansas City, we went south to Springfield and from there, east on highway 60 toward Poplar Bluff.

We stopped again in Van Buren and drove out to see the "Big Spring" we had seen advertised on billboards as we drove into town. It was every bit worth checking out. The "Big Spring" was where a swiftly flowing stream gushed out from under a huge sandstone boulder; it was a sight to see. The volume of water swirling out so fast from under the big sandstone formation was a rare sight, not just the swiftness of the stream, but the beautiful blue color of the water made it outstanding–like none other in the States, and probably in the world. I'd recommend anyone traveling that part of the country to stop and take a look.

After taking several pictures, we drove back to Van Buren for lunch and as we sat in the car having our sandwiches and cokes, two teen-aged boys who we thought were identical twins came out of the restaurant with Cokes in their hands. They sat down on the long restaurant porch dangling their feet over the side. They just sat there staring down at us and smiling. We couldn't get over how much they both looked like the kid that played the five-string banjo in the movie "Deliverance." They looked like twins, possibly inbreeds. They just sat there staring at us in the car, not saying a word, not blinking an eye.

We went from there on toward Paducah, stopping in Cairo, a city that borders three states, Kansas, Illinois, and Kentucky. Now Jeanie could add Illinois to the list of different states she'd been in. After leaving Paducah ,we traveled on toward Hopkinsville and Bowling Green and went on north toward Mammoth Cave.

On the route not far from Paducah, we stopped and took pictures of Kentucky Lake and Lake Barkley.

We had not known before that the two lakes that were close to each other had a lock system where tug boats pushed big barges into the flooded locks that when filled took the boats up where they emptied out into Lake Barkley. It was an operation like that of the Panama Canal. We never thought we would ever see anything like that on a lake in Kentucky.

Our next interesting but scary stop was the tour of a small section of the Mammoth Cave. The tour took us down through caverns to the main cavern, 300 feet below the ground. It was a real experience for the both of us. Neither of us had ever been that far underground. We had been near the Oregon Caves out of Cave Junction and the famous Carlsbad Caverns, but we had never actually toured them.

The trail or catwalk going down into the cave is made of a series of stainless steel catwalks with guardrails strung with lights and fastened to the walls of the caverns with big stainless pins. The guide would explain the different rock formations of icicle-like displays of different colors and shapes at the different levels on the way down.

One of the caverns could only be entered if you were small enough to squeeze through a narrow crack it had as an entrance. No way were we going to check that one out. Only some of the spelunkers (people who made exploring caves their hobby) and a few others who were thin enough, squeezed through the cavern and checked it out. Jeanie and I could have probably made it, but didn't even want to try.

We reached the final big cavern at bottom of the 300 foot tour and we were listening to the guide, a lady park ranger, telling us about how different rock formations were formed when all of a sudden the lights in the big cavern blinked a couple of times and went out. At first the park ranger couldn't get her flash to work, but she assured us everything would be all right.

It scared the livin' hell out of most of us. We were lucky. Standing right next to Jeanie and me was a spelunker. He was a man about sixty years old wearing striped bibbed overalls and a striped railroad engineer's cap. He quietly pulled a nice, bright flashlight from his pocket and handed it to the guide for more light while she called to find out what had caused the lights to flicker and go out. They were off just a few minutes, and the light flickered again and stayed on. At that point in the tour we sort of lost interest. In fact, after that we couldn't wait to get above ground again. No more tours in caves for us.

We left after we picked up some brochures and bought a couple pieces of green glass-like rocks or "slag" that glowed in the dark. They made good gifts for Gary, for his rock collection. We stopped at a roadside stand where they were selling dishes, and bought a cheap set of dinner plates.

By the time we got some sleep and a good rest, we had reached Cumberland Falls, just as it was getting dark. We paid our camping fee at the gate and were directed to the campground, the section for trailers and campers. The campground was practically full when we pulled in, but there were plenty of spaces with tables between the taken campsites.

All the park tables were placed under tall trees. We found an empty table and backed the Bronco in, lowered the tailgate, and I fired up the Coleman lantern and set the ice chest at end of the table. Jeanie set out paper plates and started putting the food out for us to make sandwiches.

She sliced some tomatoes and set them on the backside of the table with the lettuce and other fixin's. We had just sat down and started making our sandwiches when we noticed a little paw reach up from bench seat at the backside of the table It carefully took a slice of tomato. We sat there watching as the slices of tomato disappeared under the table.

I grabbed the flashlight that was lying on the tailgate and spotted the little masked face of a small raccoon stuffing the tomatoes in his mouth. We thought he was a cute little robber. I kept the light on him as he ran and climbed the big tree behind the table. The next day when we talked to the ranger about it, he said there were a lot of raccoons in the park and this happened quite often. He said the park ranger at the gate should have warned us.

The next day I took pictures of Cumberland Falls, and some classic post card pictures of Jeanie standing with the falls as her background. We were really having some interesting and fun times on our vacation.

It was a quick trip from Cumberland Falls to Williamsburg on Interstate 75. We crossed the old iron one-way bridge over the Cumberland into the old part of town long before it got dark. Getting there that early gave us a chance to walk around in the oldest part of town, take pictures, and see some of the places I had been telling Jeanie about since going there with Laura, Lila, and Mother two years before.

We took pictures of the courthouse and the bronze plaques on the lawn in front of it that displayed the names of famous people from Whitley County, including the first lady to serve in the Civil War. I showed Jeanie the little brick house on the street behind the courthouse where Mother and the family lived after her dad remarried.

After walking around town, we got back in the car, crossed the tracks by the old train depot and drove out toward Becks Creek Mountain on the street that went by the big mansion. The mansion was where Lila had taken a picture of Mother standing beside the "Caretaker" sign at the entrance near the big gate. The street went on over the rolling hills, through part of the Cumberland college campus, and on up a narrow road that went up Becks Creek Mountain toward the old homestead place.

We drove as far as the Jones Cemetery, and after taking a few pictures, went back through Williamsburg and on south to Jellico, Tennessee where we continued sight-seeing and taking pictures, and then we spent the night in the nice Jellico campground where there was plenty of room and hot showers.

Our next stop was the Appalachian Museum near Knoxville. We really enjoyed going through the museum. There was so much to see we really didn't have time to see it all. It was the most interesting place we stopped on our trip. It was all about the history of the mountain people living in the Appalachians. There were displays of hundreds of handmade things. And the things they made both for survival and for comfort were amazing. They were quite skillful in carving.

They made pork barrels by hollowing out sections of hardwood tree trunks two feet in diameter and three feet high. They made rocking chairs, and life-sized carvings of funny looking "mountain people" or "hillbillies" of different ages, dress and expressions. Like I said, we could have spent a week there and still not have seen it all.

There is a lot to see also when you travel south on Highway 66, which takes you through Sevierville, Pigeon Forge, and on through Gatlinburg in the Great Smoky Mountains, and on to Cherokee in North Carolina on 441.

We traveled the most scenic of highways and turnpikes where both the median strips and both sides of the highway looked like manicured park lawns. They were beautifully maintained. We saw no bottles or cans or garbage from fast food restaurants cluttering the roadsides like many of the highways in California and other states. Off on the side roads there were long sandy driveways leading to beautiful homes and little snow-white, Baptist country churches trimmed in black where Christian families were having picnics and playing horseshoes.

When we left Cherokee, we had to make up time, so we didn't stop at the first RV Park we came to, and by the time we did stop at rest stop, it was after midnight. We pulled in, used the restrooms, and went back to the car to try and get some rest. It wasn't long before we decided the bright lights and the noise of the idling big rigs along with the noise of the people slamming car doors, we couldn't get any real rest. We drove around toward the back of the rest stop, where it wasn't so noisy, and we were away from the bright lights. This was a much better place to try to get some sleep. There was only one car parked quite far away. We parked and I unloaded our camping stuff–the ice chest, Coleman lantern, a couple of cartons of sodas, and Jeanie's suitcase full of shoes and changes of clothes. I set the stuff as close to the car as I could get to keep watch on it.

While I was doing this, Jeanie straightened up our sleeping bags and fluffed up our pillows and laid my loaded pistol by my pillow where I could easily grab it if need be. She was already lying down when I rolled down the back window and closed the tailgate. I then walked up, opened the driver's side door and squeezed in the back. I locked the front doors and we lay there fully clothed on top of our sleeping bags. Finally, we were going to get some sleep.

We had just gotten comfortable and had just dozed off when an old car (what looked like an car from the midsixties, an Olds sedan) pulled in about forty feet away from us near the only light pole in the parking lot. We watched as five guys piled out. By the sound of their loud voices, it was obvious they were drunk or had been drinking. Two of them were walking faster than the others, and they were quickly walking toward the Bronco. I whispered to Jeanie to lie still as they came up to the car.

There was no way I could get out in time to see what they wanted. The two walked up to the car and tried to look in through the tinted side window. With the dim light behind them, they were unable to see inside, so they walked around to the rear of the car for a bet-ter look where they could see through the clear glass.

When they walked to the back of the car where they could see in, they looked in and saw me sitting up with a pistol in my hand and it really scared them.

They turned and ran toward the others, screaming "He's got a gun, He's got a gun. The sum bitch has a gun. Git 'n the car. Git 'n the car!" All of them jumped in the car, slammed the doors and sped away. We thought they must have thought we were in the restroom and they were going to steal our stuff. We never knew what their intentions were, and we were glad that they never stuck around long enough for us to find out.

There were a couple other cars that pulled in to stay the night after they left, so we finally started to relax and were able to sleep and the rest of the night.

The next day we took Highway 40 from Ashville to Winston Salem, through Greensboro Durham and Raleigh, and on up north through Newport News to Norfolk and Portsmouth. We both wished we had another month so that we could look around some of the historic towns we had just driven through. We managed to take a few, but not many pictures.

We spent two days in Norfolk and Portsmouth trying to locate the Croaker. We checked out the Mothball Fleet and we went to a museum located in Portsmouth near the famous red Portsmouth lightship. The museum records showed the Croaker to be on display with the Nautilus in New London. I figured it had been sold to a foreign country or scrapped. (In the late '90s, I found out that it had been moved to Buffalo, New York and was on display alongside the USS Littlerock.)

I called Mother and Tom with the disappointing news of not getting to take Jeanie aboard the boat, and I told her we were headed down the coast and would see her in a couple of days.

Our trip down the coast from Portsmouth was fun. Before we left the area, we stopped and I took a brief tour of the Newport News, a Navy ship tourist attraction. Jeanie didn't go aboard. She waited for me at the ticket gate. She didn't feel like climbing the ladders or steep steps going to the different levels of the ship, both up and down. I only checked out the main deck and took a couple pictures. After we left, we traveled south, made a short stop at Virginia Beach, and from there went on to Myrtle Beach.

Myrtle Beach was nice but very crowded, and the color of the dark beach sand looked dirty to us. It didn't look like the clean sandy beaches of the west coast of Florida and California.

We stopped along the way and took a few pictures. When we crossed into Florida, we took Interstate 10 from Jacksonville to Tallahassee, north to Havana, and on to Mother's place.

We arrived in late afternoon. Mother, Laura, Tom and his son Tommy were waiting there to greet us. Mother's house was on a nice piece of property. She had bought the property, which was a little over five acres, and Bill had bought a nice trailer, which they lived in for two years before his death. Mother was now living in the house I had been working on when I visited her on my vacations.

Laura's property had a small house, an older trailer house, and had a fair-sized storeroom for her antique lamps and furniture, those she refinished to be sold out of her antique shop in town.

Mother didn't have a bed for us, so we spent the nights at Laura's where we had a full-size bed. Laura's daughter Laura Jean, her husband Kent, and their two sons Jerry and James lived in Tallahassee. Laura Jean and her husband owned and ran a dry cleaners and laundry in town. They came out that evening after they closed their shop.

Tom, his two children, Candy and Tommy, and Bill's son Bill were living with Mother. Bill and Mother had raised both families. Tom and his wife Madelyn were now divorced. They had separated before Mother and Bill and his three children moved to Florida. Neither Madelyn nor Tom took any real responsibility in raising Candy and Tommy. Both Candy and Tommy are very grateful for having their grandmother (Granny) and Uncle Bill in those tough times.

By the time Laura Jean and family arrived, Mother and Laura had fixed us a delicious supper of fried chicken with plenty of gravy, mashed potatoes, fresh vegetables and loads of Mother's biscuits, and for dessert we had a delicious cherry cobbler pie. Later on in the evening we enjoyed some of Tom's special margaritas and cracked up listening to his jokes.

The next day, we couldn't believe what we were seeing when he showed us his pet pig.

The little pig that Tom called "Rufus" had started out with was now a big boar that romped and played all over the yard with his young son Tommy's dog. They acted like a couple of pups, chasing each other round and round with the big hog nipping at the dog, as the dog in turn would jump up and bite the big boar's ears.

The first few months, the pig would follow Tom into the house and Tom would pick him up and lay him by his side on the couch when he took a nap; but, as little pig reached his full size, big boar got just a little too big to have him inside. When there was no way Tom could get the pig to climb up and sleep by him on the couch, they decided to put him in a pen outside. When he was first kept outside, he started sleeping under the kitchen floor at the back of the house, which was a nice, cool, ventilated spot. (The homes in Florida and Georgia are built high off the ground because of the torrential rains.) He kept going farther and farther back under the house until he ended up breaking the water lines under the kitchen sink. Mother had a pen built for him, but he would root his way out after a heavy rain. There seemed no way to contain him. Both Mother and Tom became very frustrated trying to keep him in his pen, and after few months, Mother and Tom, much to their sadness, agreed to have him butchered. Rufus ended up in the big freezer on the porch.

There was a lot to see in and around Tallahassee, if Jeanie and I had the time. We did go fishing at Lake Jackson, and Laura, Mother, and Jeanie and I made a trip to Quincy to look at the old Victorian plantation houses and big tobacco barns.

The classy Victorians with their tall columns that supported the high porches or balconies were a sight to see. Many of the plantation homes and big tobacco sheds were built before or just after the Civil War and were handed down from generation to generation. Some of the sheds were now used as shelters for RVs and trailers.

I wanted to show Jeanie Wakulla Spring, but we didn't have time. I wanted Jeanie to see as many places as I could before we started back to Santa Rosa.

Wakulla Spring was a popular tourist attraction not far from Tallahassee. The big, deep, fresh water spring is located right at the edge of a swamp. Wakulla Spring water was cold and crystal clear. The water is twenty degrees cooler than the water in swamp where several signs are posted to warn tourists of the dangerous alligators they might run into that swim in the swamp close by.

Mother, my nephew Tommy and my sister Laura and I went on a tour of Wakulla Spring in a big, glass-bottom boat where we got to see a giant catfish named Henry.

Henry was the biggest catfish imaginable.

Looking down through the magnifying crystal clear water made the fish look even bigger than it was. It looked half the size of the glass-bottom boat. It was hard to believe that a catfish could get that big.

When we reached the center of the spring, the man running the boat stopped the engine. The tour guide took out a bullhorn and started calling for Henry to come and get his dinner as he "chummed up" the huge fish by throwing big pieces of bread over the side of the boat. We were all amazed when we saw the monster fish as it swam under the boat waiting to be fed.

After the fish was fed, we slowly made our way to the spot where the spring began.

When we looked down through the crystal clear water to the bottom of the spring, we could easily see the large cave where the water was swirling out. The guide told us how far the cave went back and about the discovery of a skull and some bones of a saber-toothed tiger found by two divers exploring the cave.

The day after the trip to Wakulla Springs, we went north to Cordele, Georgia to visit Lila and her son Bill and his family.

The trip home was pretty tiring. Although we passed some pretty interesting places, the trip going home wasn't nearly as much fun. We stopped in Alabama and rested in the campground of an historic site. It had been a relay station or point from which freight wagons hauled loads of cannon balls to the battlefields for the Confederate Army.

We stopped in Nashville, Tennessee to have the electrical system in the Bronco checked out. We had to replace the flasher and two stop light bulbs. We stopped along the way at the different little restaurants and enjoyed the delicious southern cooked food like that we grew up on. The menus in the little places had our kind of food. We loved the fried chicken, biscuits and gravy and fried green tomatoes and cornbread and beans and turnip greens served on nice big plates in the little country restaurants. There was no place we could find good food like that in restaurants west of the Rockies.

We had planned on stopping in Arkansas and visiting one of Jeanie's cousins, but again we didn't have time. We traveled Highway 40 out of Memphis, which took us by Fort Smith, Arkansas, Oklahoma City, Amarillo, Texas, and on to Albuquerque, and from there we took 666 southwest of Gallup to see the northern part of the Petrified Forest. We wanted to travel further south and see more, but didn't have the time. We took a few pictures, went back up, got back on Highway 40, drove to Kingman, took some pictures of Hoover Dam, went on to Las Vegas, and on home.

We had a few thrills on the way back. One night as we were going down a long hill on a real curvy section of highway, we thought we were going to be run over by the big rigs hauling freight through the area. Big rigs with blinding headlights were towing three trailers. They were like short trains tailgating us all the way. We would find a turnout, pull off the road to let them pass and just as we got back on the road, another one would be right on our bumper again.

It was miserably hot when we reached Las Vegas. I wanted to get a motel for the night and have the air conditioner looked at the next morning, but after talking it over, we decided to wait and have it fixed when we got home. We made it home without any engine problems and had the air conditioner recharged soon after.

Jeanie was anxious to get home to check on her garden and geese, anxious to see if she had any new goslings or baby chicks that may have hatched out while we were gone.

The garden and fruit trees and grapevines needed a good watering, but otherwise, everything else was doing just fine. We had paid her son Bill and his two children to water the birds and garden and flower plants while we were gone.

The geese were especially happy to have Jeanie home. She had names for her geese and other birds, and kept a logbook of those sitting on eggs to be hatched. One whole section of the backyard, beyond where she had the garden, was filled with various pens and cages for her birds, and we had another pen for two wild hogs. At one time she had quite a variety of birds. Some she had bought at the feed store, and she was helping raise some pheasant chicks for one of the guys I worked with. She had pens of several kinds of chickens and geese both domestic and wild, and also some guineas that weren't penned. They were the watchdogs of yard and sounded off when any stranger or strange dog came into the yard. They had the run of the property, but hung out in the backyard.

10 A GOOD LIFE

In the late eighties, our sons Jim and Gary were married and it wasn't too long before we were blessed with three not only beautiful, but also very intelligent, grandchildren.

On April 26, 1990, Debbie gave birth to our first grandchild, James Patrick Creekmore III, and a little over a year later we were again blessed with the birth of our first and only granddaughter, Kelli Nichole Creekmore, Jimmy's little sister. Jeanie and I were so proud of the two beautiful grandchildren with their big brown eyes. We always looked forward to babysitting them. About a year later we were so delighted again when Julie, Gary's wife, gave birth to our third grandchild, Louis Nathan Creekmore. I don't think there were grandparents anywhere in the world that enjoyed their grandchildren any more than we did.

Jeanie and I were so happy when we got to babysit them on weekends. We were always fixing things for them and playing with them. Jeanie always fixed their favorite foods and always had special movies for them to watch. One movie they never seemed to tire of watching while having their breakfast was Jungle Book. Little Kelli always seemed to find something to occupy her. She was very creative. In the kitchen she would rummage through the stuff in the catchall drawer by the sink and build little make-believe things she would put together on the kitchen table. They all three enjoyed the swing I had made for them in the backyard and the toy bows and arrows, and Kelli often played in the sandbox I had made for her.

Even during the difficult times, they certainly took our minds off the problems we were going through.

Just before I retired from UPS, the company changed medical companies from Health Plan of the Redwoods to Kaiser Permanente.

We had no idea what serious troubles we were in for concerning our medical problems. Debbie, our daughter-in law, tried to warn us. She told us about the serious blunders and mistakes that Kaiser doctors in the Bay Area and other parts of the state had made. Mistakes that not only left patients crippled or mutilated when they removed the wrong arm or leg, but even operations that resulted in unnecessary deaths.

At first we were real pleased with the hospital. We liked the convenience of being able to be treated so close by for any kind of illness. The hospital complex was new and had just about everything patients needed for a medical problem.

With the exception of some x-ray equipment and MRI machines, everything was right there in the complex. There were lists of doctors, including specialists that handled every kind of an injury or medical problem.

But in the early summer of 1991, shortly before I retired, things started to get pretty rough, and I guess I should say depressing, for our family.

Not only were we worried about Mother's health, but we had also become very concerned about Jeanie. We found out that she was having heart problems along with serious breathing problems. I had been taking her back and forth to a Kaiser lung specialist and two heart specialists. After several appointments and x-rays with the heart specialists, we were told that she would eventually have to have open-heart surgery. X-rays and tests showed that she had to have not only one, but two heart valves replaced.

I was really frustrated to say the least. I had been flying back and forth to Tallahassee to check on Mother who was in and out of the hospital with heart and breathing problems, too.

Everything seemed to be going wrong. In midsummer of 1991, the neighbor's dogs killed most of Jeanie's special silky chickens, several ducks, and her pride and joy, her two Canadian geese, her two favorite pets. What a thing to happen on top of everything else that was going on in our lives.

If I had gotten up that morning before daylight and walked out back to the chicken pen when I first heard a high-pitched whine from one of the neighbor's dogs, I might have been able to do something about it. When I heard the noise, I just thought the neighbor had gotten a new pup to add to her collection of flea-bags.

The whine mingled with the sounds of the cars on the busy freeway nearby. It sounded like it was coming from a pup or a small dog. We had no idea that the sound came from the biggest dog of the pack, which was quietly mauling Jeanie's most treasured birds.

I had to get up before six to go to work, so I got up, closed the window to block out the noise of the whining dog, and went back to bed, burying my head in the pillow. If only I had walked to the backyard when I heard the first shrill dog noises, I might have prevented the slaughter.

The filthy, mangy dogs had threaded their way through the piles of debris and dog crap in the neigh-bor's yard and had managed to squeeze under the fence to maul and kill Jeanie's ducks, chickens and our prize Canadian Honker geese.

In the mornings when the neighbor got up to take her stroll in the backyard, it was no surprise to see three or four dogs of different sizes and breeds come charging out the back door to romp and play and do their business, urinating and leaving big piles of you-know-what in the driveway between our two houses and sometimes in our backyard, if they could manage to squeeze under the fence.

She was divorced and had a house full of fleabag dogs and cats and a cage with a couple of birds. We had no idea what we were in for when she and her family moved in just about a year after I had started paying on our place. If we had known what a dirty neighbor we were getting next door, we would never have bought the place.

After we found out how trashy they were, Jeanie and I started looking for another place, but we couldn't find a place with as much land that was also within reasonable driving distance of the Christian school we had our boys in.

As the saying goes, you can pick your friends but not your neighbors.

Later that morning when I walked back to the chicken, goose, and duck pens, my heart went to my shoe tops. There before me were both of Jeanie's prize Canadian geese, lying on the ground in the bottom of the pen covered in blood. There was a big hole in the pen that I had built for them.

It was quite obvious from the hair hanging on the wire around the hole at the pen's bottom that this is where the big dog had forced his way through. I had tried several times to talk to the neighbor about hooking her dogs up to a dog run like those we had kept our dogs on when we had them. The poor geese living in the closed pen had nowhere go to get away from the dog. It was likely the mangy retriever had been the one that had mauled them to death while the other black and white female dog was killing the other birds and Jeanie's little Silky chickens, all the while making her irritating, high-pitched sounds when she couldn't break into some of the cages she was pawing at, trying get to the birds inside.

When I went back to the house, Jeanie was coming out the back door. I stopped her and had her sit down before I told her what had happened. Jeanie and I both were heartbroken. Now we were not only heart-broken over Mother being so sick, but overwhelmed with the loss of her birds. I confronted the neighbor, but she had the nerve to look me straight in the eye and say her dogs had nothing to do with the death of the birds. She said they had been in her house when the mauling took place.

Several months later, her son admitted to me that he and his Mother had been back near the chicken pens that morning and had seen her dogs go on their killing rampage from pen to pen. They had calmed the dogs and taken them inside.

The next day, son Jim came up for a visit and was furious when we told him what had happened. That same day, the golden retriever came back through the fence and Jim and I caught him and hauled him off. We put him in the back of my pickup and hauled him about forty miles south and let him out in the country not far from Dillon Beach in Marin County.

I wanted to shoot him and leave him for the buzzards, but Jim talked me out of it. He said the dog wouldn't last long if he went near a chicken house in one of the farmyards.

Three months went by and we thought the dog was dead and gone until the day I was coming back from the mailbox and saw the neighbor pull in her driveway with that worthless, scroungy dog in the back of her car.

I wished I had gone ahead and killed the dog when we first hauled it off. I knew she had posters out to notify her if the dog was found. She had retrieved the dog after getting a call from a dog pound in Marin County. Jeanie and I often wondered about her mentality. How could a Berkeley college graduate be so lacking in common sense? We wondered how anyone could be so stupid as to do the idiotic things she did.

That fall I had another confrontation with our good neighbor. I had to call the health department and turn her in for dumping kitty litter in the weeds all along the fence line. The idiot said she dumped it there to keep the dogs from coming into our yard. We didn't notice it until after the first heavy rain when the white granules were flooding Jeanie's raspberries and grapevines.

♪

And then things got even worse. I took early retirement in 1991 so I could be with Jeannie as her health continued to deteriorate. Then, the good doctors at Kaiser suggested that I have prostate surgery even though my PSA levels were 0.3–within what I later discovered is considered a normal range. I insisted on radiation treatments instead because there was no way I could be lifting my wife in and out of bed if I had surgery, even though she was very small. I was taking care of her after her surgery, which we felt had been delayed much longer than it should have been. I'll never ever forget the worry and heartaches this brought on the family. I wished there was someway I could get even with Kaiser Permanente for these mistakes, but suing them would be like shovelin' sand against the tide. It wouldn't bring her back, anyway–Jeannie passed away July 17th, on our oldest son's birthday.

A man doesn't realize what a woman goes through till he takes on her job. I could not believe a woman the size of Jeanie could flop over a mattress on a full size bed and stretch on a fitted sheet. And changing a sheet was a minor part of all the jobs she took on around our little farm. She raised big vegetable gardens and canned a big part of the produce, and she kept our big freezer in the garage filled with frozen salmon and ling cod that I had caught on my fishing trips on the *Jeanie II*.

Regardless of all the hard work she did during the day, she always looked like she just came back from the beauty salon when she served me my dinner. Her out-look on life was always very positive. I would give anything to have her by my side so we could finish out our time in this old world together. I just try to think of all the good times we shared throughout our courtship and marriage. To me, she was my special angel.

♪

Those years, even with all the heartaches we went through were good years, the best years of our lives. We were very proud of both of our sons Jim and Gary. After working security jobs in Bodega Bay, patrolling the roads that wind up and down the hills between the expensive homes near and around the big golf course, they both ended up with jobs in law enforcement. It's hard for me to believe they are grown up now.

Jim, our oldest son had worked for the campus police at Santa Rosa Jr. College and went through the Police Academy in Sonoma. He went to work in 1984 for the San Pablo Police Department and became the acting Chief of Police in San Pablo, retiring after twenty eight years of service. Gary worked as a Corrections Officer at a state prison near Safford, Arizona and spent time training recruit corrections officers in Tucson. Gary and his wife Julie are both now working long, stressful hours at the federal prison in Eloy, Arizona. Gary is working in Control Count and Julie is working as Unit Manager.

The two grandsons, Jimmy and Louis, became Marines, Jimmy completing two tours in Afghanistan and returning home safely; Louis was deployed to Afghanistan also, as part of Marine Corps Special Operations involved in light-armored recon.

My granddaughter Kelli has just finished her college courses in nursing.

I often look back on all the fond memories and the wonderful times I spent with my sons, their wives, and our grandchildren.

What's next for me? I'm relatively healthy for 81 years old, and I'm sure there are adventures ahead. I plan to teach the great grandchildren how to play the mandolin, if they're interested, and I've begun to write about the wild side of UPS delivery. So who knows? There may be more travelin' in my future.

TRAVELIN' WITH THE POOR BOY FROM BECKS CREEK

Aunt Mildred on the homestead, Becks Creek, KY

Creekmore Kids, Cripple Creek, CO

Racy "business cards"

List of Photographs (credits unknown)

p 11	Permeile, Bumper, Mother
p. 35	James P. Creekmore, Penrose, 1940
p. 115	Leta and Laura Creekmore, Canutillo, TX, 1947
p 182	USS Zelima
p 184	USS Zelima Shipmates
p 212	Parthenon 1956
p 221	USS Croaker near Malta
p 226	Monaco Souvenir
p 230	Gibralter 1956
p 258	Jeanie Creekmore
p 272	Car Wreck
p 283	Jeanie & First Son Gary
p 366	Aunt Mildred on the homestead, Becks Creek, KY
	The Creekmore Kids, Cripple Creek, CO
	Racy "business cards"

ABOUT THE AUTHOR

James P. Creekmore lives in Redmond, Oregon. This is his first book.

Made in the USA
San Bernardino, CA
17 March 2017